The Bold Theories

Shah Rukh

Published by Shah Rukh, 2024.

While every precaution has been taken in the preparation of this book, the publisher assumes no responsibility for errors or omissions, or for damages resulting from the use of the information contained herein.

THE BOLD THEORIES

First edition. May 18, 2024.

Copyright © 2024 Shah Rukh.

Written by Shah Rukh.

Table of Contents

Introduction

In the vast expanse of human knowledge, there have always been pioneers who dared to challenge the boundaries of conventional wisdom, pushing the frontiers of understanding to new and uncharted territories. These bold thinkers, armed with curiosity, creativity, and a relentless pursuit of truth, have forged groundbreaking theories that have revolutionized our understanding of the universe and reshaped the course of human history.

"The Bold Theories" is a journey into the realm of revolutionary ideas and paradigm-shifting concepts that have transformed our perception of reality. From the depths of quantum mechanics to the expanses of cosmology, this book explores the visionary minds and daring hypotheses that have propelled humanity forward on the path of scientific discovery.

Through the pages of this book, readers will embark on a captivating exploration of some of the most audacious theories ever conceived. From the revolutionary insights of Albert Einstein and his theory of relativity to the enigmatic realms of quantum mechanics and the mysteries of the multiverse, "The Bold Theories" delves into the minds of the visionaries who dared to challenge the status quo and redefine our understanding of the cosmos.

Each chapter of this book offers a compelling narrative that illuminates the genesis, development, and impact of a seminal theory, inviting readers to embark on a voyage of discovery through the annals of scientific history. From the elegant simplicity of Newton's laws to the mind-bending complexities of string theory, "The Bold Theories" invites readers to explore the rich tapestry of human intellect and imagination.

As we venture into the realm of bold theories, we confront not only the mysteries of the universe but also the boundless potential of the human mind. For it is through the courage to question, the audacity to dream, and the willingness to embrace the unknown that we unlock the secrets of the cosmos and chart a course towards a deeper understanding of our place in the universe.

Join us on this extraordinary journey as we delve into the realm of "The Bold Theories" and witness the transformative power of human ingenuity in the pursuit of knowledge and truth.

Chapter 1: Simulation Hypothesis

The Simulation Hypothesis, also known as the Simulation Theory or Simulation Argument, proposes that our reality, the universe, and everything within it, including all matter, energy, and life forms, could potentially be an artificial simulation or computer-generated construct. This idea has gained traction in both scientific and philosophical circles, sparking intense debate and speculation about the nature of reality and our place within it. In this detailed exploration, we'll delve into the origins, proponents, arguments, criticisms, and implications of the Simulation Hypothesis.

Origins and Background:

The concept of reality as a simulation is not a recent phenomenon. It can be traced back to ancient philosophical inquiries about the nature of existence, such as Plato's Allegory of the Cave, which posits that what we perceive as reality may merely be shadows cast by a higher truth. However, the modern articulation of the Simulation Hypothesis gained prominence with the rapid advancements in computer technology and the increasing sophistication of video games and virtual reality simulations.

One of the earliest explicit formulations of the Simulation Hypothesis is attributed to the philosopher Nick Bostrom in his seminal paper "Are You Living in a Computer Simulation?" published in 2003. Bostrom's argument is based on statistical reasoning and posits that at least one of the following propositions must be true:

1. Human civilization is unlikely to reach a posthuman stage.
2. Posthuman civilizations are not interested in running ancestor simulations.
3. We are almost certainly living in a computer simulation.

Proponents and Arguments:

Numerous proponents from various fields, including philosophy, science, and technology, have explored and supported the Simulation Hypothesis. Elon Musk, the CEO of SpaceX and Tesla, has famously expressed his belief in the likelihood of our reality being a simulation, suggesting that the odds of us living in "base reality" are one in billions. Musk's argument is based on the assumption that technological advancements will inevitably lead to the creation of highly realistic simulations indistinguishable from physical reality.

Other proponents, such as physicist Neil deGrasse Tyson and philosopher David Chalmers, while not explicitly endorsing the Simulation Hypothesis, have acknowledged its plausibility and the importance of considering such speculative ideas in the pursuit of understanding the nature of reality. They argue that as our computational abilities continue to improve, we may eventually possess the capability to create entire simulated worlds inhabited by conscious entities.

Arguments and Evidence:

Proponents of the Simulation Hypothesis often cite several lines of reasoning and evidence to support their claims:

1. **Computational Power**: The exponential growth of computational power and the development of advanced artificial intelligence suggest that it may eventually become possible to simulate entire universes with sentient beings.
2. **Anthropic Principle**: The Anthropic Principle, which states that the universe must be compatible with the existence of observers (since we are here to observe it), can be interpreted as evidence in favor of the Simulation Hypothesis. If our reality is simulated, it would naturally be finely tuned to

support the emergence of conscious life forms.

3. **Quantum Phenomena**: Some proponents argue that certain quantum phenomena, such as wave-particle duality and the observer effect, resemble characteristics of computer simulations, where events are computed only when observed.

4. **Simulation Paradox**: The lack of evidence against the Simulation Hypothesis is often framed as evidence in its favor. If we were living in a simulated reality, it would be logically consistent for the creators to prevent us from discovering the true nature of our existence.

Criticisms and Challenges:

Despite its intriguing implications, the Simulation Hypothesis has faced a fair share of criticisms and challenges:

1. **Empirical Evidence**: Critics argue that there is a lack of empirical evidence to support the Simulation Hypothesis. While analogies to computer simulations and philosophical arguments may be thought-provoking, they do not constitute concrete proof of a simulated reality.

2. **Occam's Razor**: Some critics invoke Occam's Razor, the principle that the simplest explanation is usually the best, to argue against the Simulation Hypothesis. They contend that positing the existence of a complex simulation is unnecessary when simpler explanations, such as the natural laws governing our universe, suffice.

3. **Ethical Considerations**: The ethical implications of creating or inhabiting simulated realities raise complex moral questions. If we were to simulate conscious beings, would they have rights and moral standing? What responsibilities would the creators of such simulations have toward their simulated inhabitants?

4. **Infinite Regress**: Accepting the Simulation Hypothesis leads to the possibility of an infinite regress of simulations within simulations, each containing simulated realities of their own. This raises the question of where the chain of simulations began and whether there is a "base reality" outside of all simulations.

Implications and Speculations:

The Simulation Hypothesis, if true, would have profound implications for our understanding of reality, consciousness, and existence:

1. **Nature of Reality**: If our reality is indeed a simulation, it would fundamentally alter our perception of what is real and what is simulated. The distinction between physical and virtual reality would blur, challenging traditional notions of existence.

2. **Consciousness**: The Simulation Hypothesis raises intriguing questions about the nature and origin of consciousness. If consciousness can emerge within simulated environments, it suggests that consciousness may not be an exclusively biological phenomenon but rather a fundamental aspect of reality.

3. **Existential Questions**: The realization that we might be living in a simulated reality prompts existential questions about the purpose and meaning of our existence. If our lives are part of a simulation, what is the ultimate goal or intention behind its creation?

4. **Technological Ramifications**: Exploring the Simulation Hypothesis could drive advancements in fields such as artificial intelligence, virtual reality, and computational neuroscience. It could also influence our approach to the development of future technologies and our understanding of

the universe as a vast computational system.

Conclusion:

The Simulation Hypothesis offers a provocative lens through which to examine the nature of reality, consciousness, and existence. While it remains a speculative hypothesis with no definitive proof, its exploration has sparked interdisciplinary dialogue and prompted profound questions about the fabric of reality. Whether our reality is ultimately revealed to be a simulation or not, contemplating the possibility challenges us to reconsider our assumptions and expand our understanding of the universe and our place within it.

Chapter 2: The Holographic Principle

The Holographic Principle is a fascinating concept in theoretical physics that proposes a deep connection between gravity, quantum mechanics, and the structure of the universe. In this detailed exploration, we'll delve into the origins, development, implications, and ongoing research surrounding the Holographic Principle.

Origins and Background:

The concept of the Holographic Principle emerged from efforts to reconcile the principles of general relativity, which describes gravity on large scales, with those of quantum mechanics, which govern the behavior of particles on the smallest scales. It finds its roots in the study of black holes and the information paradox.

The idea gained significant traction in the late 20th century, particularly through the work of physicist Gerard 't Hooft and, later, Leonard Susskind. 't Hooft proposed that the information about the contents of a black hole could be encoded on its event horizon, akin to a holographic projection. Susskind further developed this notion, suggesting that the entire universe might be viewed as a hologram, with all the information about its contents encoded on its boundary.

The Principle:

At its core, the Holographic Principle posits that the information describing a three-dimensional volume of space can be fully represented on a two-dimensional surface bounding that volume. This implies that the contents of a region of space can be entirely described by data on its boundary, much like a holographic image, where a three-dimensional scene is encoded on a two-dimensional surface.

Mathematically, this principle is often expressed through the analogy of the AdS/CFT correspondence, which relates certain gravitational theories in a higher-dimensional Anti-de Sitter space (AdS) to conformal field theories (CFTs) on the boundary of that space. According to this correspondence, the physics within the AdS space can be fully described by the physics of the boundary CFT.

Implications and Applications:

The Holographic Principle has profound implications for our understanding of fundamental physics and the nature of spacetime:

1. **Black Hole Physics**: The Holographic Principle offers insights into the information paradox of black holes. Instead of suggesting that information is lost within a black hole's event horizon, it implies that all the information about the objects consumed by a black hole is encoded on its surface. This has led to fruitful avenues of research in black hole thermodynamics and quantum gravity.

2. **Quantum Gravity**: By providing a framework for understanding the quantum properties of gravity, the Holographic Principle has spurred advancements in the quest for a theory of quantum gravity. It suggests that the fundamental building blocks of spacetime may emerge from more fundamental quantum degrees of freedom residing on lower-dimensional surfaces.

3. **Cosmology**: The holographic nature of the universe implies that the entire cosmos can be thought of as a projection from a lower-dimensional boundary. This perspective offers new avenues for understanding cosmological phenomena, such as the nature of the Big Bang, the evolution of the universe, and the cosmic microwave background radiation.

4. **Quantum Information Theory**: The Holographic Principle

has profound implications for quantum information theory. It suggests that the amount of information in a given volume of space is limited by its surface area, rather than its volume. This has implications for quantum computing, cryptography, and our understanding of entropy.

Development and Research:

Since its inception, the Holographic Principle has been a fertile ground for theoretical exploration and experimental validation:

1. **String Theory**: The Holographic Principle finds a natural home within string theory, a theoretical framework that posits that the fundamental constituents of the universe are one-dimensional "strings." String theory provides a mathematical framework for understanding the holographic nature of spacetime through the AdS/CFT correspondence.
2. **Quantum Gravity Research**: Physicists continue to investigate the implications of the Holographic Principle for theories of quantum gravity, such as loop quantum gravity, causal dynamical triangulation, and string theory. These efforts aim to reconcile general relativity and quantum mechanics at the most fundamental level.
3. **Experimental Tests**: While direct experimental tests of the Holographic Principle are challenging due to the extreme scales involved, researchers have proposed various indirect tests and observational signatures. These include studying the properties of black holes, cosmic microwave background radiation, and high-energy particle physics experiments.
4. **Cosmological Observations**: Observational cosmology provides indirect evidence for the holographic nature of the universe. Measurements of the cosmic microwave background and large-scale structure formation can provide insights into

the underlying quantum information structure of spacetime.

Criticisms and Challenges:

Despite its theoretical elegance and explanatory power, the Holographic Principle is not without its critics and challenges:

1. **Mathematical Rigor**: Some critics argue that the mathematical formalism underlying the Holographic Principle is still not fully understood and lacks the necessary rigor. The AdS/CFT correspondence, while a powerful tool, has limitations and may not capture the full complexity of quantum gravity.
2. **Information Paradox**: The resolution of the black hole information paradox within the framework of the Holographic Principle remains an active area of research. While the principle suggests that information is conserved and encoded on black hole horizons, the details of how this information is retrieved remain elusive.
3. **Experimental Verification**: Direct experimental verification of the Holographic Principle is challenging due to the extreme conditions involved. As a result, much of the evidence for the principle comes from theoretical arguments, mathematical consistency, and indirect observational signatures.
4. **Interpretational Issues**: The holographic nature of reality raises profound interpretational issues about the nature of spacetime, information, and the fundamental constituents of the universe. These issues touch upon deep questions about the nature of reality and our place within it.

Conclusion:

The Holographic Principle represents a profound insight into the nature of spacetime, quantum mechanics, and gravity. While still a speculative idea, it has inspired generations of physicists and mathematicians to explore new avenues of research and deepen our understanding of the fundamental structure of the universe. Whether it ultimately proves to be a fundamental principle of nature or a mathematical curiosity, the Holographic Principle continues to shape our exploration of the deepest mysteries of the cosmos.

Chapter 3: Multiverse Theory

The Multiverse Theory is a concept that extends beyond the boundaries of our observable universe, proposing the existence of multiple universes, each with its own unique set of physical laws, constants, and conditions. This expansive idea has captivated the imaginations of physicists, cosmologists, philosophers, and science fiction enthusiasts alike, prompting extensive exploration into its theoretical underpinnings, implications, and potential methods of verification. In this comprehensive exploration, we'll delve into the origins, different types, supporting evidence, criticisms, and philosophical implications of the Multiverse Theory.

Origins and Background:

The concept of a multiverse has its roots in ancient philosophical and religious traditions, where ideas of parallel worlds or alternate realities have been contemplated for centuries. However, the modern scientific exploration of the multiverse began with developments in theoretical physics and cosmology in the 20th century.

One of the earliest formulations of the multiverse concept arose from quantum mechanics. The Many-Worlds Interpretation, proposed by physicist Hugh Everett III in 1957, posited that every quantum event results in the universe splitting into multiple branches, each representing a different outcome. This interpretation implies the existence of a vast ensemble of parallel universes, collectively forming the multiverse.

Types of Multiverse:

The Multiverse Theory encompasses various proposed mechanisms for generating multiple universes. Some of the most prominent types include:

1. **Many-Worlds Interpretation**: As mentioned earlier, in the Many-Worlds Interpretation of quantum mechanics, every quantum event spawns a branching universe, resulting in a branching tree of parallel realities.

2. **Inflationary Multiverse**: Inflationary cosmology suggests that the early universe underwent a period of rapid expansion, or inflation. Within this framework, regions of space could undergo separate inflationary phases, leading to the formation of "bubble" universes within a larger multiverse.

3. **Brane Multiverse**: In string theory, which posits that fundamental particles are actually tiny, vibrating strings, the universe may exist as a "brane" embedded in a higher-dimensional space. Collisions or interactions between these branes could give rise to separate universes, each with its own physical properties.

4. **Parallel Universes in Quantum Mechanics**: Certain interpretations of quantum mechanics, such as the Many-Worlds Interpretation, imply the existence of parallel universes where different quantum outcomes are realized.

5. **Ultimate Multiverse**: Some theories, such as Max Tegmark's Mathematical Universe Hypothesis, propose that all mathematically consistent structures exist as separate universes within a vast mathematical landscape.

Supporting Evidence:

While direct observational evidence for the multiverse remains elusive, several lines of reasoning and theoretical considerations support the concept:

1. **Cosmic Microwave Background**: Observations of the cosmic microwave background radiation, the afterglow of the Big Bang, have revealed anomalies and patterns that some

researchers suggest could be indicative of a larger, multiverse-scale structure.

2. **String Theory Predictions**: String theory, a theoretical framework attempting to unify quantum mechanics and general relativity, predicts the existence of multiple dimensions and parallel universes. While string theory has yet to be experimentally verified, its mathematical consistency lends credence to the possibility of a multiverse.

3. **Fine-Tuning of Physical Constants**: The remarkable fine-tuning of the fundamental physical constants of our universe, such as the strength of gravity and the cosmological constant, has led some scientists to propose that our universe may be just one among many, each with different values for these constants.

4. **Inflationary Cosmology**: The theory of cosmic inflation, which explains the homogeneity and isotropy of the observable universe, naturally leads to the idea of a multiverse containing regions with different properties that resulted from variations in the initial conditions of inflation.

5. **Quantum Mechanics and Many-Worlds**: While controversial, the Many-Worlds Interpretation of quantum mechanics provides a theoretical basis for the existence of parallel universes, where every quantum event results in the creation of multiple branches of reality.

Criticisms and Challenges:

Despite its theoretical appeal, the concept of the multiverse is not without its critics and challenges:

1. **Falsifiability**: Many versions of the multiverse theory are difficult, if not impossible, to test experimentally, raising concerns about its status as a scientific hypothesis rather than

a speculative idea.

2. **Occam's Razor**: The principle of Occam's Razor, which favors simpler explanations over more complex ones, leads some critics to question the necessity of invoking the multiverse to explain observed phenomena when simpler explanations may suffice.

3. **Philosophical and Metaphysical Concerns**: The multiverse raises profound philosophical questions about the nature of reality, existence, and our place within the cosmos. Critics argue that the concept may verge into metaphysical speculation rather than empirical science.

4. **Anthropic Principle**: The anthropic principle, which states that the universe must be compatible with the existence of observers (since we are here to observe it), can be interpreted as both supporting and challenging the multiverse. While the multiverse could explain the apparent fine-tuning of our universe, it also raises questions about the validity of making statistical inferences across an infinite number of universes.

5. **Lack of Direct Evidence**: Perhaps the most significant criticism of the multiverse is the lack of direct observational evidence. While theoretical arguments and mathematical consistency provide some support for the idea, the inability to directly observe or test other universes remains a significant hurdle.

Philosophical Implications:

The concept of the multiverse has far-reaching philosophical implications that extend beyond the realm of physics:

1. **Nature of Reality**: The existence of a multiverse challenges traditional conceptions of reality, suggesting that our universe may be just one among many, each with its own set of physical

laws and conditions.

2. **Copernican Principle**: The multiverse reinforces the Copernican principle, which asserts that humanity does not occupy a privileged position in the cosmos. Instead, we may be just one of countless civilizations scattered across an infinite multiverse.

3. **Ethical and Existential Questions**: The multiverse raises profound ethical and existential questions about the nature of existence, free will, and the meaning of life. If there are infinitely many versions of ourselves in parallel universes, what does that mean for concepts such as personal identity and moral responsibility?

4. **Limits of Knowledge**: The concept of the multiverse highlights the limits of human knowledge and the inherent uncertainty of our understanding of the cosmos. It underscores the need for humility and open-mindedness in our pursuit of scientific and philosophical truth.

Conclusion:

The Multiverse Theory represents a bold and ambitious attempt to understand the fundamental nature of reality and our place within it. While still a speculative idea with many unresolved questions and challenges, it has sparked intense debate and exploration across multiple disciplines, from theoretical physics to philosophy to cosmology. Whether the multiverse ultimately proves to be a scientific reality or a philosophical speculation, its exploration pushes the boundaries of human knowledge and our understanding of the universe.

Chapter 4: Panspermia

Panspermia is a hypothesis that suggests life exists throughout the universe and can be spread from one planetary system to another via comets, asteroids, or other means of space travel. The term "panspermia" comes from the Greek words "pan," meaning "all," and "sperma," meaning "seed," implying that life is ubiquitous and can be disseminated across cosmic distances. In this comprehensive exploration, we will delve into the origins, different types, supporting evidence, criticisms, and implications of the Panspermia hypothesis.

Origins and Background:

The idea of panspermia has ancient roots, with early philosophers and thinkers pondering the possibility of life traveling through space. However, the modern concept of panspermia gained traction in the 19th century with the work of scientists such as Lord Kelvin and Hermann von Helmholtz. Kelvin proposed the idea of "cosmic dust" carrying life between planets, while Helmholtz speculated about microbial life being transported through space.

The term "panspermia" was popularized by the Swedish chemist Svante Arrhenius in the early 20th century. Arrhenius proposed that microbial life could be propelled through space by radiation pressure from stars, eventually seeding other planets with life.

Types of Panspermia:

Panspermia encompasses various mechanisms by which life could be transferred between celestial bodies. Some of the key types include:

1. **Lithopanspermia**: In lithopanspermia, life is transported within rocks or meteoroids. Microorganisms could survive the harsh conditions of space within these protective

structures and potentially colonize other planets upon impact.

2. **Radiopanspermia**: Radiopanspermia proposes that life could be carried on radiation-resistant organisms, such as extremophiles, which could survive the journey through space and subsequent exposure to cosmic radiation.

3. **Directed Panspermia**: Directed panspermia is a more speculative idea, suggesting that intelligent beings could intentionally seed life on other planets. This concept was famously proposed by the Nobel laureate Francis Crick and chemist Leslie Orgel in the 1970s.

4. **Interstellar Panspermia**: Interstellar panspermia involves the transfer of life between star systems. This could occur through the ejection of material from a planet's surface due to impact events, or through the gravitational influence of passing stars.

Supporting Evidence:

While direct evidence for panspermia remains elusive, several lines of reasoning and observations lend credence to the hypothesis:

1. **Extremophiles**: The discovery of extremophiles—organisms capable of surviving extreme conditions such as high radiation, vacuum, and temperature fluctuations—suggests that life may be more resilient and adaptable than previously thought, bolstering the idea that it could survive space travel.

2. **Meteorite Contamination**: Studies of meteorites have revealed organic compounds and even traces of microbial life. While some of these findings are controversial and subject to debate, they provide tantalizing hints that life may be capable of hitching a ride on space rocks.

3. **Space Missions**: Experiments conducted on space missions, such as NASA's EXPOSE project, have demonstrated that certain microorganisms can survive prolonged exposure to the

vacuum and radiation of space. These findings support the idea that life could potentially endure the journey between planets.

4. **Interplanetary Transfer**: The discovery of Martian meteorites on Earth has led to speculation about the possibility of life originating on Mars and being transported to Earth via meteorite impacts. While the evidence remains inconclusive, it raises intriguing questions about the potential interplanetary transfer of life.

Criticisms and Challenges:

Despite its appeal, panspermia also faces several criticisms and challenges:

1. **Survival in Space**: While some experiments suggest that certain microorganisms can survive the conditions of space, the harsh environment of interplanetary travel poses significant challenges to the survival of life. Critics argue that the likelihood of organisms surviving the journey intact is extremely low.

2. **Origin of Life**: Panspermia does not address the ultimate origin of life but rather the transportation mechanism between celestial bodies. Critics argue that it merely shifts the question of life's origins to another location, without providing a satisfactory explanation.

3. **Probability**: The probability of panspermia events occurring and successfully seeding life on habitable planets is uncertain and difficult to quantify. Critics contend that invoking panspermia as an explanation for the origin of life may be overly speculative without concrete evidence.

4. **Directed Panspermia**: The concept of directed panspermia, while intriguing, raises ethical and philosophical questions

about the intentions and motivations of hypothetical extraterrestrial beings. Critics argue that the idea is highly speculative and lacks empirical support.

Implications and Significance:

Panspermia has profound implications for our understanding of the origin and distribution of life in the universe:

1. **Extraterrestrial Life**: Panspermia suggests that life may be more widespread in the universe than previously thought. If life can be transported between planets, it could potentially increase the likelihood of finding extraterrestrial life within our own solar system or beyond.
2. **Astrobiology**: The study of panspermia has implications for astrobiology, the search for life beyond Earth. Understanding the potential mechanisms by which life could be transported between celestial bodies informs our strategies for exploring other planets and moons for signs of life.
3. **Planetary Protection**: Panspermia also has implications for planetary protection and the prevention of biological contamination during space exploration. If life can survive the journey between planets, there is a risk of inadvertently contaminating pristine environments with terrestrial organisms.
4. **Origin of Life**: While panspermia does not provide a definitive answer to the origin of life, it offers an alternative perspective that complements other theories, such as abiogenesis and directed evolution. It suggests that life may have originated elsewhere in the universe and been transported to Earth, potentially influencing the development of life on our planet.

Conclusion:

Panspermia is a captivating hypothesis that challenges our traditional notions of the origin and distribution of life in the universe. While still speculative and subject to debate, it offers a plausible mechanism by which life could be disseminated across cosmic distances, potentially influencing the development of life on habitable planets. As our understanding of astrobiology and space exploration continues to advance, panspermia remains a topic of ongoing research and speculation, with implications that extend far beyond the boundaries of our own planet.

Chapter 5: The Gaia Hypothesis

The Gaia Hypothesis, also known as Gaia Theory, proposes that Earth is a self-regulating, living organism capable of maintaining conditions conducive to life. This idea, initially formulated by chemist James Lovelock and biologist Lynn Margulis in the 1970s, challenges traditional views of Earth as an inert, inanimate planet and suggests that life plays an active role in shaping its environment. In this detailed exploration, we will delve into the origins, key principles, supporting evidence, criticisms, and implications of the Gaia Hypothesis.

Origins and Background:

The Gaia Hypothesis takes its name from Gaia, the ancient Greek goddess of the Earth. The idea emerged from interdisciplinary discussions between Lovelock and Margulis, who sought to understand the interconnectedness of Earth's biosphere, atmosphere, oceans, and geology. Drawing inspiration from systems theory, cybernetics, and evolutionary biology, they proposed that Earth functions as a complex, self-regulating system akin to a living organism.

Key Principles:

The Gaia Hypothesis is built on several key principles:

1. **Homeostasis**: At the heart of the Gaia Hypothesis is the concept of homeostasis, the ability of living organisms to maintain stable internal conditions despite external changes. Earth, according to the hypothesis, regulates its environment through feedback mechanisms that stabilize temperature, composition, and other factors necessary for life.
2. **Feedback Mechanisms**: Gaia operates through a network of feedback loops, both positive and negative, that regulate

environmental conditions. Positive feedback amplifies changes, while negative feedback counteracts them, maintaining equilibrium. Examples include the carbon cycle, temperature regulation, and the regulation of atmospheric oxygen levels.

3. **Holism**: The Gaia Hypothesis emphasizes the interconnectedness of all living and non-living components of Earth's system. It views Earth as a single, integrated organism, with each component influencing and being influenced by the whole.

4. **Evolutionary Perspective**: Gaia is not static but dynamic, evolving over time in response to changes in the environment and the evolution of life. The hypothesis suggests that life and its environment co-evolve, with organisms shaping their surroundings and vice versa.

Supporting Evidence:

While the Gaia Hypothesis remains controversial, several lines of evidence and observations lend support to its key principles:

1. **Biogeochemical Cycles**: The Earth's biogeochemical cycles, such as the carbon, nitrogen, and phosphorus cycles, exhibit characteristics of self-regulation. Feedback mechanisms within these cycles help maintain stable conditions necessary for life.

2. **Climate Regulation**: Life plays a crucial role in regulating Earth's climate through processes such as photosynthesis, respiration, and the release of greenhouse gases. The presence of life has helped stabilize Earth's temperature over geological timescales, preventing runaway heating or cooling.

3. **Atmospheric Composition**: The composition of Earth's atmosphere, particularly the presence of oxygen and methane,

is largely determined by biological processes. These gases, in turn, regulate the climate and support the existence of life.

4. **Planetary Boundaries**: The concept of planetary boundaries, introduced by scientists in the 2000s, aligns with the principles of the Gaia Hypothesis. It identifies critical thresholds for various Earth systems, such as biodiversity loss, climate change, and ocean acidification, beyond which irreversible environmental changes could occur.

Criticisms and Challenges:

Despite its appeal, the Gaia Hypothesis has faced criticism and skepticism from some scientists:

1. **Teleological Interpretation**: Critics argue that the Gaia Hypothesis can be misinterpreted as teleological, implying that Earth has a purpose or goal. They caution against anthropomorphizing Earth or attributing agency to the planet itself, as opposed to the interactions of living organisms.

2. **Lack of Predictive Power**: Some critics contend that the Gaia Hypothesis lacks predictive power and testability, making it difficult to validate empirically. While the hypothesis offers a compelling narrative for understanding Earth's systems, it remains challenging to apply in a predictive or quantitative manner.

3. **Selection Bias**: The Gaia Hypothesis may suffer from selection bias, as proponents may focus on examples that support the hypothesis while overlooking evidence to the contrary. Critics argue that Earth's history contains periods of environmental instability and mass extinctions inconsistent with the notion of a self-regulating Gaia.

4. **Reductionist Critique**: The Gaia Hypothesis has been

criticized for its rejection of reductionism, the idea that complex phenomena can be understood by analyzing their constituent parts. Critics argue that focusing solely on holistic, emergent properties may overlook important underlying mechanisms and processes.

Implications and Significance:

The Gaia Hypothesis has profound implications for our understanding of Earth's systems and our relationship with the planet:

1. **Environmental Awareness**: The Gaia Hypothesis fosters a deeper appreciation for the interconnectedness of life on Earth and the fragility of the planet's systems. It underscores the importance of stewardship and sustainable management of natural resources.
2. **Planetary Health**: Viewing Earth as a self-regulating system highlights the importance of maintaining the health and resilience of its ecosystems. Human activities that disrupt natural processes, such as deforestation, pollution, and climate change, can destabilize Earth's systems and threaten the stability of Gaia.
3. **Astrobiology**: The Gaia Hypothesis has implications for the study of astrobiology and the search for life beyond Earth. It suggests that the presence of life on a planet may influence its habitability and the detectability of biosignatures from space.
4. **Philosophical Reflections**: The Gaia Hypothesis invites philosophical reflections on humanity's place in the cosmos and our responsibility as stewards of Earth. It challenges anthropocentric views of the planet and calls for a more holistic and ecological perspective on our relationship with nature.

Conclusion:

The Gaia Hypothesis offers a compelling framework for understanding Earth as a dynamic, self-regulating system teeming with life. While controversial and subject to debate, it has sparked interdisciplinary discussions and influenced thinking in fields ranging from ecology and biology to philosophy and environmental science. Whether viewed as a scientific theory, metaphor, or heuristic device, the Gaia Hypothesis invites us to reconsider our relationship with the planet and to recognize the interconnectedness of all life on Earth.

Chapter 6: Riemann Hypothesis

The Riemann Hypothesis is one of the most famous and enduring unsolved problems in mathematics. Proposed by the German mathematician Bernhard Riemann in 1859, it deals with the distribution of prime numbers and has profound implications for number theory and other areas of mathematics. In this detailed exploration, we will delve into the historical context, formulation, significance, attempts at proof, and broader implications of the Riemann Hypothesis.

Historical Context:

The study of prime numbers, those integers greater than one that are divisible only by themselves and one, has fascinated mathematicians for millennia. In the early 19th century, the German mathematician Carl Friedrich Gauss made groundbreaking discoveries about the distribution of primes, laying the groundwork for subsequent investigations.

In 1859, Bernhard Riemann presented a paper titled "On the Number of Prime Numbers Less Than a Given Magnitude" to the Berlin Academy, in which he introduced the Riemann Hypothesis as a conjecture about the distribution of zeros of the Riemann zeta function.

Formulation:

The Riemann Hypothesis is often stated as follows:

"The nontrivial zeros of the Riemann zeta function all lie on the critical line with real part 1/2."

Here, the Riemann zeta function $\zeta(s)$ is a complex function defined for complex numbers s with real part greater than one. Its nontrivial zeros are the values of s for which $\zeta(s)$ equals zero and lie off the critical line with real part 1/2.

Significance:

The Riemann Hypothesis holds significant importance in number theory and related areas of mathematics for several reasons:

1. **Prime Number Theorem**: A proof of the Riemann Hypothesis would provide a rigorous justification for the Prime Number Theorem, which describes the asymptotic distribution of prime numbers. The theorem states that the number of primes less than or equal to a given number x is approximately x/log(x), a result that is closely tied to the behavior of the zeros of the Riemann zeta function.

2. **Distribution of Prime Numbers**: The Riemann Hypothesis offers insights into the distribution of prime numbers and their spacing. It predicts that primes are "equally distributed" in the sense that the gaps between consecutive primes are as small as possible, a property known as "Riemann's staircase."

3. **Connection to Other Fields**: The Riemann Hypothesis has connections to a wide range of mathematical areas, including algebraic number theory, analytic number theory, harmonic analysis, and mathematical physics. Its resolution could lead to breakthroughs in these fields and deepen our understanding of fundamental mathematical concepts.

4. **Computational Applications**: While the Riemann Hypothesis is a purely theoretical conjecture, its implications extend to practical applications in cryptography, computer science, and data security. The distribution of prime numbers plays a crucial role in cryptographic algorithms, and a proof

of the Riemann Hypothesis could have implications for the security of cryptographic systems.

Attempts at Proof:

Over the past century and a half, numerous mathematicians have attempted to prove or disprove the Riemann Hypothesis, making it one of the most actively pursued problems in mathematics. Various approaches and techniques have been employed, including:

1. **Analytic Methods**: Many attempts at proving the Riemann Hypothesis have involved analytical techniques from complex analysis, harmonic analysis, and functional analysis. These methods seek to understand the behavior of the Riemann zeta function and its zeros in the complex plane.

2. **Number Theoretic Methods**: Number theorists have developed specialized techniques for studying the distribution of primes and analyzing properties of the Riemann zeta function. These methods often involve deep results from algebraic number theory, such as class field theory and modular forms.

3. **Probabilistic Methods**: Some mathematicians have approached the Riemann Hypothesis using probabilistic techniques, treating the distribution of primes as a random phenomenon and analyzing statistical properties of prime number sequences.

4. **Computational Approaches**: With the advent of powerful computers, computational methods have become increasingly important in the study of the Riemann Hypothesis. Computer simulations and numerical experiments can provide valuable insights into the behavior of the Riemann zeta function and its zeros.

Broader Implications:

The Riemann Hypothesis has far-reaching implications beyond the realm of pure mathematics:

1. **Foundations of Mathematics**: A proof of the Riemann Hypothesis would have profound implications for the foundations of mathematics, providing insights into the structure of numbers, the nature of mathematical truth, and the limits of human knowledge.
2. **Mathematical Philosophy**: The Riemann Hypothesis raises philosophical questions about the nature of mathematical objects, the role of conjectures and proofs in mathematical reasoning, and the relationship between mathematics and reality.
3. **Technological Applications**: While the Riemann Hypothesis is primarily of theoretical interest, its resolution could have practical implications for technology, including cryptography, data encryption, and information security.
4. **Cultural Impact**: The Riemann Hypothesis has captured the imagination of mathematicians and non-mathematicians alike, inspiring countless books, articles, and documentaries. Its resolution would be a major event in the history of mathematics and would likely attract widespread media attention.

Conclusion:

The Riemann Hypothesis stands as one of the most profound and elusive problems in mathematics. Its resolution would not only provide a deeper understanding of the distribution of prime numbers but would also have far-reaching implications for number theory, mathematical philosophy, and technology. While progress has been made over the years, the Riemann Hypothesis remains one of the great unsolved mysteries of mathematics, challenging the ingenuity and creativity of generations of mathematicians to come.

32

Chapter 7: Directed Panspermia

Directed Panspermia is a speculative hypothesis that proposes the intentional seeding of life on other planets or celestial bodies by advanced extraterrestrial civilizations. Unlike the more general concept of panspermia, which suggests that life can spread through natural processes such as meteorite impacts, Directed Panspermia posits deliberate intervention by intelligent beings to initiate or accelerate the development of life elsewhere in the universe. In this comprehensive exploration, we will delve into the origins, proponents, criticisms, potential methods, and implications of Directed Panspermia.

Origins and Background:

The concept of Directed Panspermia was first proposed by the Nobel laureate Francis Crick, along with chemist Leslie Orgel, in a 1973 paper titled "Directed Panspermia." Crick, who is best known for his role in discovering the structure of DNA, speculated that life on Earth might have been intentionally seeded by an extraterrestrial civilization billions of years ago. This idea emerged in the context of ongoing debates about the origin of life and the possibility of extraterrestrial intelligence.

Proponents:

Directed Panspermia has been championed by a small but vocal group of scientists, philosophers, and science fiction authors who find the idea intriguing and worthy of consideration. Some notable proponents include:

1. **Francis Crick**: As one of the original proponents of Directed Panspermia, Crick argued that the complexity of life on Earth and the rapid emergence of biological diversity could be

explained by intentional intervention from an advanced extraterrestrial civilization.

2. **Leslie Orgel**: Orgel collaborated with Crick on the Directed Panspermia hypothesis and contributed to the development of the concept. He suggested that life could be transported through space on spacecraft or other means of interstellar travel.

3. **Carl Sagan**: While not a staunch advocate of Directed Panspermia, the renowned astronomer Carl Sagan speculated about the possibility of life spreading through space and advocated for further research into the topic.

Potential Methods:

Directed Panspermia proposes various methods by which intelligent beings could seed life on other planets or celestial bodies:

1. **Spacecraft**: One possibility is that advanced civilizations could send robotic spacecraft or probes containing microbial life to target destinations within their own solar system or beyond. This spacecraft could be equipped with the necessary technology to survive the journey through space and initiate the process of colonization upon arrival.

2. **Genetic Information**: Rather than sending physical organisms, another approach is to transmit genetic information encoded in digital or biological form. This information could contain the instructions for synthesizing life forms once they reach their destination, similar to the concept of "digital DNA."

3. **Astroengineering**: In addition to seeding life directly, advanced civilizations could engage in large-scale astroengineering projects to terraform planets or moons and make them more hospitable to life. This could involve altering

the atmosphere, temperature, and other environmental conditions to create habitats suitable for colonization.

Criticisms and Challenges:

Directed Panspermia remains a speculative hypothesis with many unanswered questions and criticisms:

1. **Evidence**: One of the primary criticisms of Directed Panspermia is the lack of empirical evidence to support the hypothesis. While it is theoretically possible that life could be intentionally seeded by extraterrestrial civilizations, there is currently no direct evidence to confirm this.
2. **Fermi Paradox**: The Fermi Paradox, which asks why we have not detected any signs of extraterrestrial intelligence despite the vastness of the universe, poses a challenge to Directed Panspermia. If advanced civilizations exist, why have they not made their presence known or communicated with us?
3. **Occam's Razor**: Some critics argue that Directed Panspermia violates the principle of Occam's Razor, which favors simpler explanations over more complex ones. They contend that the hypothesis introduces unnecessary assumptions about the origin of life and the existence of extraterrestrial civilizations.
4. **Ethical Concerns**: The idea of intentionally seeding life on other planets raises ethical questions about the rights and responsibilities of intelligent beings to manipulate the development of life in the universe. Critics caution against potential unintended consequences and the risk of ecological disruption.

Implications and Significance:

Directed Panspermia has profound implications for our understanding of the origin and evolution of life in the universe:

1. **Origin of Life**: If Directed Panspermia were true, it would provide a novel explanation for the origin of life on Earth and potentially other planets. Instead of arising spontaneously through abiogenesis, life could have been deliberately seeded by extraterrestrial beings.

2. **Astrobiology**: The hypothesis of Directed Panspermia has implications for the field of astrobiology, the study of life in the universe. It suggests that life may be more widespread and diverse than previously thought, with potential implications for the search for extraterrestrial life.

3. **Exoplanet Exploration**: Directed Panspermia raises questions about the ethics and implications of future space exploration and colonization efforts. If life can be intentionally seeded on other planets, what are the implications for our own efforts to explore and colonize other worlds?

4. **Philosophical Reflections**: The idea of Directed Panspermia invites philosophical reflections on the nature of life, intelligence, and the role of humanity in the cosmos. It raises questions about our place in the universe and our relationship to other forms of intelligent life.

Conclusion:

Directed Panspermia is a speculative hypothesis that proposes the intentional seeding of life on other planets or celestial bodies by advanced extraterrestrial civilizations. While controversial and subject to criticism, it has captured the imagination of scientists, philosophers, and science fiction enthusiasts alike. Whether or not Directed

Panspermia is ultimately confirmed, it serves as a thought-provoking exploration of the possibilities and implications of life in the universe.

37

Chapter 8: Hollow Earth Theory

The Hollow Earth Theory is a speculative hypothesis proposing that the Earth is not a solid sphere but instead contains hollow regions or cavities within its interior. This idea has captivated the imaginations of explorers, scientists, and conspiracy theorists for centuries, inspiring numerous works of fiction and fueling debates about the nature of our planet. In this comprehensive exploration, we will delve into the historical origins, proponents, evidence (or lack thereof), criticisms, and cultural significance of the Hollow Earth Theory.

Historical Origins:

The concept of a hollow Earth has ancient roots, with myths and legends from cultures around the world describing subterranean realms inhabited by gods, spirits, or mythical creatures. However, the modern incarnation of the Hollow Earth Theory emerged during the Scientific Revolution of the 17th and 18th centuries, fueled by advances in astronomy, geology, and exploration.

One of the earliest proponents of the Hollow Earth Theory was the English astronomer and mathematician Edmond Halley, best known for calculating the orbit of the comet that bears his name. In 1692, Halley proposed that the Earth might consist of concentric shells, with openings at the poles through which sunlight could enter, illuminating the interior.

Proponents:

Over the centuries, the Hollow Earth Theory has attracted a diverse array of proponents, including scientists, explorers, and fringe theorists. Some notable proponents include:

1. **John Cleves Symmes Jr.:** In the early 19th century, Symmes,

an American army officer, became one of the most fervent advocates of the Hollow Earth Theory. He proposed that the Earth was composed of a series of concentric spheres, each with its own inhabited surface and polar openings.

2. **Cyrus Reed Teed**: Teed, a 19th-century American physician and self-proclaimed prophet, espoused a variant of the Hollow Earth Theory known as the "Concave Hollow Earth" or "Cellular Cosmogony." He believed that we live on the inside of a hollow sphere, with the universe contained within.

3. **Richard E. Byrd**: The American naval officer and explorer Richard E. Byrd is sometimes cited as evidence for the Hollow Earth Theory due to his purported expeditions to the North and South Poles. While Byrd's actual expeditions were scientific in nature and did not involve discovering a hollow Earth, his name is often invoked by proponents of the theory.

Evidence (or Lack Thereof):

Despite its enduring popularity, the Hollow Earth Theory lacks empirical evidence and is widely regarded as pseudoscience by the scientific community. Critics point to several lines of evidence that contradict the theory:

1. **Gravity**: One of the most fundamental objections to the Hollow Earth Theory is the principle of gravity. According to our understanding of physics, a hollow Earth would not be able to maintain its spherical shape without collapsing under its own gravitational force.

2. **Seismic Waves**: The study of seismic waves generated by earthquakes and other geological phenomena provides strong evidence for the Earth's internal structure. Seismic data indicate that the Earth has a solid, layered interior consisting of a crust, mantle, and core, rather than being hollow.

3. **Geological Features**: The Earth's surface features, such as mountains, valleys, and ocean trenches, provide further evidence against the Hollow Earth Theory. These features are consistent with the processes of plate tectonics and continental drift, which are well-understood by geologists.

4. **Polar Exploration**: Despite claims to the contrary, no credible evidence has ever been presented to support the existence of polar openings or vast hollow regions at the Earth's poles. Arctic and Antarctic expeditions have thoroughly explored these regions and found no indication of entrances to a hollow Earth.

Criticisms:

The Hollow Earth Theory has faced criticism from scientists, skeptics, and scholars for its lack of empirical evidence, reliance on pseudoscientific reasoning, and promotion of conspiracy theories. Some of the key criticisms include:

1. **Pseudoscience**: Critics argue that the Hollow Earth Theory lacks scientific rigor and is based on flawed assumptions, speculative reasoning, and pseudoscientific principles. The theory fails to meet the criteria of falsifiability, testability, and empirical verification required by mainstream science.

2. **Conspiracy Theories**: The Hollow Earth Theory has been associated with various conspiracy theories, including claims of government cover-ups and secret expeditions to explore the Earth's interior. These conspiracy theories often rely on anecdotal evidence, hearsay, and misinformation to support their claims.

3. **Confirmation Bias**: Proponents of the Hollow Earth Theory often exhibit confirmation bias, selectively interpreting evidence to support their preconceived beliefs while ignoring

or dismissing contradictory evidence. This cherry-picking of data undermines the credibility of the theory and hinders scientific progress.

4. **Cultural Influences**: The popularity of the Hollow Earth Theory can be attributed in part to its appeal as a narrative trope in literature, film, and popular culture. While entertaining as a work of fiction, the theory lacks substantive scientific merit and should be approached with skepticism.

Cultural Significance:

Despite its status as pseudoscience, the Hollow Earth Theory continues to hold cultural significance as a source of inspiration for writers, artists, and filmmakers. The idea of a hidden world beneath our feet has captured the imagination of generations and spawned countless works of fiction, including novels, films, and television shows.

1. **Literature**: The Hollow Earth motif has been a recurring theme in literature, appearing in works such as Jules Verne's "Journey to the Center of the Earth," Edgar Rice Burroughs' "Pellucidar" series, and H.G. Wells' "The War of the Worlds." These stories often depict fantastical subterranean worlds populated by strange creatures and civilizations.

2. **Film and Television**: The Hollow Earth concept has been popularized in film and television, with numerous movies and TV shows featuring explorers journeying into the Earth's interior. Examples include the film "The Core," the TV series "Lost," and the animated film "Ice Age: Dawn of the Dinosaurs."

3. **Conspiracy Culture**: The Hollow Earth Theory has also found a niche within conspiracy culture, where it is sometimes cited as evidence for government cover-ups and secret knowledge. Conspiracy theorists often invoke the Hollow

Earth as part of larger narratives about hidden truths and forbidden knowledge.

Conclusion:

The Hollow Earth Theory is a speculative hypothesis proposing that the Earth contains hollow regions or cavities within its interior. Despite its lack of empirical evidence and scientific credibility, the theory has captured the imagination of explorers, scientists, and conspiracy theorists for centuries. While entertaining as a work of fiction, the Hollow Earth Theory should be approached with skepticism and critical thinking, as it lacks substantive scientific support. Nevertheless, its cultural significance as a source of inspiration for literature, film, and popular culture continues to endure.

Chapter 9: Fermi Paradox

The Fermi Paradox is a fascinating and enduring puzzle in the field of astrophysics and the search for extraterrestrial intelligence (SETI). Named after the Italian-American physicist Enrico Fermi, the paradox refers to the apparent contradiction between the high probability of the existence of extraterrestrial civilizations and the lack of evidence for, or contact with, such civilizations. In this detailed exploration, we will delve into the historical context, formulation, proposed solutions, criticisms, and implications of the Fermi Paradox.

Historical Context:

The Fermi Paradox originated from a casual conversation between Enrico Fermi and colleagues during a lunch at the Los Alamos National Laboratory in 1950. The discussion centered on the possibility of interstellar travel and the likelihood of extraterrestrial civilizations. Fermi famously posed the question, "Where is everybody?" reflecting the puzzling absence of evidence for the existence of advanced extraterrestrial civilizations in the observable universe.

Formulation:

The Fermi Paradox can be succinctly stated as follows:

1. **High Probability of Existence**: Given the vast number of stars and planets in the observable universe, along with the age of the universe, the probability of the existence of technologically advanced extraterrestrial civilizations capable of interstellar communication or travel is considered to be high.
2. **Lack of Evidence**: Despite the high probability of their existence, we have not detected any definitive evidence of

extraterrestrial civilizations, such as signals from advanced technological societies or physical artifacts of their presence.

Proposed Solutions:

Over the years, scientists, philosophers, and scholars have proposed various solutions to the Fermi Paradox, each offering possible explanations for the apparent absence of extraterrestrial civilizations. Some of the key solutions include:

1. **Rare Earth Hypothesis**: The Rare Earth Hypothesis suggests that Earth-like planets capable of supporting complex life may be rare in the universe. Factors such as the presence of a stable star, a terrestrial planet within the habitable zone, and a protective magnetic field could all contribute to the rarity of intelligent life.
2. **Great Filter Hypothesis**: The Great Filter Hypothesis posits that there may be one or more improbable steps in the process of the emergence and evolution of intelligent life that act as a "filter," preventing the development of advanced civilizations. If this filter lies in the past, it would explain why we have not encountered extraterrestrial civilizations.
3. **Technological Singularity**: Some theorists propose that advanced civilizations may undergo a technological singularity, a hypothetical future event where technological progress accelerates to the point of surpassing human comprehension. Such civilizations may transcend the need for physical communication or exploration, rendering them undetectable to us.
4. **Extraterrestrial Isolationism**: The Extraterrestrial Isolationism hypothesis suggests that advanced civilizations may deliberately avoid contact with other civilizations, either out of fear, a desire to preserve their own culture and

autonomy, or adherence to a principle of non-interference.

5. **Zoo Hypothesis**: The Zoo Hypothesis posits that extraterrestrial civilizations may be aware of our existence but have chosen to observe us without interference, akin to a zookeeper observing animals in a zoo. They may be waiting for humanity to reach a certain level of development before making contact.

Criticisms:

Despite the proposed solutions, the Fermi Paradox remains a topic of ongoing debate and speculation. Some of the key criticisms and challenges include:

1. **Anthropocentric Bias**: Critics argue that the Fermi Paradox may be influenced by anthropocentric bias, the tendency to view the universe from a human-centric perspective. Our assumptions about the behavior and motivations of extraterrestrial civilizations may be limited by our own experiences and understanding.

2. **Limits of Detection**: Our current methods of searching for extraterrestrial intelligence, such as radio telescopes and optical surveys, may be limited in their ability to detect signals or artifacts of extraterrestrial civilizations. Advanced civilizations may use communication methods or technologies beyond our current capabilities.

3. **Temporal Asymmetry**: The Fermi Paradox assumes that extraterrestrial civilizations would have emerged and developed at roughly the same time as humanity. However, if civilizations arise at different points in time, they may not overlap temporally, reducing the likelihood of contact.

4. **Sample Size**: The search for extraterrestrial intelligence has only explored a tiny fraction of the observable universe,

leaving vast regions unexplored. The lack of evidence for extraterrestrial civilizations may simply be a result of the limited scope of our observations.

Implications:

The Fermi Paradox has profound implications for our understanding of the universe and our place within it:

1. **Cosmic Loneliness**: The Fermi Paradox raises existential questions about humanity's place in the cosmos and our sense of cosmic loneliness. If we are truly alone in the universe, it has implications for our understanding of life, intelligence, and the purpose of existence.
2. **Technological Progress**: The search for extraterrestrial intelligence has driven advances in astronomy, astrophysics, and technology, leading to new methods and techniques for exploring the universe. Regardless of the outcome, the pursuit of understanding our cosmic neighbors has expanded our knowledge and capabilities.
3. **Future Exploration**: The Fermi Paradox serves as a motivation for future space exploration and the search for extraterrestrial life. Whether we find evidence of extraterrestrial civilizations or not, the quest to explore the universe and unlock its mysteries will continue to inspire generations of scientists and explorers.
4. **Ethical Considerations**: The Fermi Paradox also raises ethical considerations about our potential interactions with extraterrestrial civilizations. If we were to detect evidence of intelligent life elsewhere in the universe, how should we approach communication, cooperation, and potential conflicts?

Conclusion:

The Fermi Paradox remains one of the most intriguing and enduring mysteries in science, challenging our assumptions about the prevalence and behavior of extraterrestrial civilizations. While numerous solutions have been proposed, the paradox continues to defy easy explanation, serving as a reminder of the vastness and complexity of the universe. As our understanding of the cosmos deepens and our technology advances, the search for answers to the Fermi Paradox will undoubtedly continue to drive scientific inquiry and exploration for generations to come.

Chapter 10: Grand Unified Theory (GUT)

The Grand Unified Theory (GUT) is a theoretical framework in particle physics that aims to unify three of the four fundamental forces of nature: electromagnetism, the weak nuclear force, and the strong nuclear force. The ultimate goal of GUT is to provide a single, comprehensive explanation for the fundamental interactions of particles and fields, thereby achieving a deeper understanding of the fundamental structure of the universe. In this detailed exploration, we will delve into the historical context, key principles, experimental evidence, challenges, and implications of the Grand Unified Theory.

Historical Context:

The quest for a Grand Unified Theory dates back to the early 20th century, with the formulation of quantum mechanics and the development of quantum field theory. The unification of electromagnetism and the weak nuclear force was achieved in the 1960s with the formulation of the electroweak theory by Sheldon Glashow, Abdus Salam, and Steven Weinberg, which predicted the existence of the W and Z bosons.

Building on these advances, physicists sought to extend the unification to include the strong nuclear force, leading to the development of Grand Unified Theories in the 1970s and beyond. While GUTs have not yet been confirmed experimentally, they remain a topic of active research and speculation in theoretical physics.

Key Principles:

The Grand Unified Theory is based on several key principles and concepts:

1. **Gauge Symmetry**: GUTs are typically formulated within the framework of gauge field theory, which describes the interactions of particles and fields in terms of local symmetry transformations. Gauge symmetries dictate the form of the fundamental forces and the particles that mediate them.

2. **Unified Gauge Group**: In a Grand Unified Theory, the electromagnetic, weak, and strong forces are described by a single, unified gauge group. This group is typically a simple Lie group such as $SU(5)$, $SO(10)$, or $E(6)$, which contains the symmetries of all three forces.

3. **Spontaneous Symmetry Breaking**: GUTs often involve the phenomenon of spontaneous symmetry breaking, whereby the symmetry of the unified gauge group is broken at high energies, giving rise to the distinct forces observed at lower energies. This process predicts the existence of new particles and interactions.

4. **Unification Scale**: GUTs predict the existence of a unification scale, typically around 10^{15} to 10^{16} GeV, at which the three forces become indistinguishable and merge into a single unified force. This scale is much higher than the energies accessible to current particle accelerators.

Experimental Evidence:

While GUTs remain speculative and have not yet been confirmed experimentally, there are several lines of indirect evidence and theoretical motivations supporting the idea of grand unification:

1. **Quantum Numbers**: The electroweak theory successfully unified the electromagnetic and weak forces by introducing a new quantum number known as weak isospin. This suggests that a similar approach may be possible for unifying the strong force as well.

2. **Mass Hierarchy**: The masses of elementary particles exhibit a hierarchy, with the masses of the W and Z bosons being much larger than those of the photon and gluons. This hierarchy is suggestive of a deeper underlying symmetry that could be revealed at higher energies.

3. **Proton Decay**: Many GUTs predict that protons, the stable building blocks of atomic nuclei, can decay into lighter particles with lifetimes on the order of 10^30 to 10^32 years. While no proton decay events have been observed to date, ongoing experiments such as Super-Kamiokande and others continue to search for evidence of this process.

4. **Neutrino Oscillations**: The phenomenon of neutrino oscillations, in which neutrinos change flavor as they propagate through space, provides indirect evidence for physics beyond the Standard Model. GUTs often incorporate mechanisms to explain the masses and mixing of neutrinos.

Challenges:

Despite the theoretical appeal of Grand Unified Theories, several challenges and open questions remain:

1. **Experimental Verification**: The most significant challenge facing GUTs is the lack of experimental evidence. The energies required to probe the unification scale are far beyond the capabilities of current particle accelerators, making direct tests of GUT predictions difficult.

2. **Proton Stability**: The predicted lifetime of the proton in GUTs varies depending on the specific model, but all predictions exceed the current experimental limits by many orders of magnitude. The absence of observed proton decay events poses a significant challenge to GUTs.

3. **Neutrino Masses**: While GUTs can accommodate the masses

and mixing of neutrinos, the specific mechanisms responsible for generating neutrino masses remain poorly understood. Experimental measurements of neutrino properties, such as their absolute masses and the nature of their mass hierarchy, could provide important clues.

4. **Cosmological Observations**: Cosmological observations, such as the cosmic microwave background radiation and the large-scale structure of the universe, provide valuable constraints on GUT models. Any viable GUT must be consistent with the observed properties of the universe on large scales.

Implications:

The successful formulation and experimental confirmation of a Grand Unified Theory would have profound implications for our understanding of the universe:

1. **Unified Description of Forces**: GUTs aim to provide a unified description of the fundamental forces of nature, revealing the underlying symmetry and structure of the universe. This would constitute a major triumph of theoretical physics and a significant step toward a "theory of everything."

2. **Origin of Mass**: GUTs may shed light on the origin of particle masses and the mechanism responsible for electroweak symmetry breaking. Understanding why particles have the masses they do is one of the central mysteries of particle physics.

3. **Cosmological Significance**: GUTs have implications for the early universe and the processes that occurred shortly after the Big Bang. They may help explain the origin of matter-antimatter asymmetry, the generation of primordial fluctuations, and other cosmological puzzles.

4. **Experimental Technologies**: The search for evidence of grand unification has driven advances in experimental techniques and technologies, including particle accelerators, detectors, and cosmological observatories. These advances benefit not only particle physics but also other fields of science and technology.

Conclusion:

The Grand Unified Theory represents a bold and ambitious endeavor to unify three of the four fundamental forces of nature within a single theoretical framework. While GUTs remain speculative and have not yet been confirmed experimentally, they offer a compelling vision of a deeper underlying unity in the fabric of the universe. As theoretical and experimental physics continue to advance, the quest for grand unification remains a central goal of modern particle physics, promising to deepen our understanding of the fundamental structure and dynamics of the cosmos.

Chapter 11: Boltzmann Brain

The Boltzmann Brain is a provocative and controversial concept in theoretical physics and cosmology that challenges our understanding of the universe and the nature of reality. Named after the Austrian physicist Ludwig Boltzmann, the Boltzmann Brain hypothesis posits that it is more probable for a spontaneously fluctuating conscious entity, such as a brain, to appear in the universe than for the universe as a whole to exist in its observed form. In this detailed exploration, we will delve into the historical context, formulation, criticisms, potential implications, and philosophical ramifications of the Boltzmann Brain hypothesis.

Historical Context:

The origins of the Boltzmann Brain concept can be traced back to Ludwig Boltzmann's work in statistical mechanics and thermodynamics in the late 19th century. Boltzmann developed the concept of entropy, a measure of disorder or randomness in a system, and formulated statistical laws describing the behavior of large ensembles of particles. His work laid the foundation for modern statistical physics and provided insights into the fundamental principles governing the behavior of matter and energy.

The Boltzmann Brain hypothesis emerged in the context of debates about the ultimate fate of the universe, the nature of cosmological constants, and the implications of quantum mechanics and cosmology for our understanding of reality. It represents a radical departure from conventional cosmological models and challenges our intuitions about the nature of existence.

Formulation:

The Boltzmann Brain hypothesis can be formulated as follows:

1. **Statistical Fluctuations**: According to the laws of quantum mechanics and statistical mechanics, the universe is subject to random fluctuations at the microscopic level. These fluctuations can give rise to temporary deviations from the average state of the universe, including the spontaneous appearance of conscious entities such as brains.

2. **Thermodynamic Considerations**: From a thermodynamic perspective, the universe tends toward states of maximum entropy, or disorder. Over time, the universe will evolve toward a state of maximum entropy, known as heat death, in which all energy is uniformly distributed and no further thermodynamic work can be done.

3. **Boltzmann Brains vs. Observable Universe**: The Boltzmann Brain hypothesis suggests that it is more probable for a Boltzmann Brain—a spontaneously fluctuating conscious entity—to appear in the universe than for the universe as a whole to exist in its observed form. This is because the probability of a small fluctuation giving rise to a Boltzmann Brain is higher than the probability of the entire universe spontaneously fluctuating into existence.

Criticisms:

The Boltzmann Brain hypothesis has been met with skepticism and criticism from physicists, cosmologists, and philosophers. Some of the key criticisms include:

1. **Violation of Occam's Razor**: Critics argue that the Boltzmann Brain hypothesis violates Occam's razor, the principle of parsimony, which favors simpler explanations over more complex ones. The hypothesis posits the existence of highly improbable, self-aware entities arising from random fluctuations, which some view as less plausible than alternative

explanations for the nature of the universe.

2. **Fine-Tuning Problems**: The Boltzmann Brain hypothesis raises fine-tuning problems related to the observed values of cosmological constants and parameters. In order for Boltzmann Brains to be the dominant form of conscious entities in the universe, the values of fundamental constants such as the cosmological constant and the vacuum energy density would need to be vastly different from their observed values, leading to inconsistencies with observational data.

3. **Cosmological Observations**: Observational evidence from cosmology, such as the cosmic microwave background radiation, the large-scale structure of the universe, and the observed distribution of galaxies, provides constraints on cosmological models and the nature of the universe. The predictions of the Boltzmann Brain hypothesis may be inconsistent with these observations.

4. **Anthropic Considerations**: The Boltzmann Brain hypothesis relies on anthropic reasoning, which takes into account the conditions necessary for the existence of observers like ourselves. However, anthropic arguments can be controversial and lead to counterintuitive conclusions if not carefully applied.

Potential Implications:

Despite its controversial nature, the Boltzmann Brain hypothesis raises thought-provoking questions and potential implications for our understanding of the universe:

1. **Nature of Consciousness**: The hypothesis challenges our understanding of consciousness and its relationship to the physical universe. If consciousness can arise from random fluctuations in the structure of the universe, it suggests a

deeply interconnected relationship between the physical world and subjective experience.

2. **Multiverse Scenarios**: The Boltzmann Brain hypothesis has implications for multiverse scenarios, which posit the existence of multiple universes with different physical properties. In some multiverse models, the appearance of Boltzmann Brains may be more common than in others, leading to different predictions for the nature of reality.

3. **Cosmological Evolution**: The hypothesis raises questions about the long-term evolution and fate of the universe. If the universe is subject to random fluctuations and thermal equilibrium, it may undergo cycles of expansion, contraction, and re-emergence, with conscious entities such as Boltzmann Brains appearing sporadically throughout cosmic history.

4. **Philosophical Considerations**: The Boltzmann Brain hypothesis has philosophical implications for our understanding of the nature of existence, the role of consciousness in the universe, and the limits of scientific knowledge. It challenges traditional metaphysical assumptions and invites us to reexamine our conceptions of reality.

Conclusion:

The Boltzmann Brain hypothesis represents a bold and provocative idea at the intersection of physics, cosmology, and philosophy. While controversial and subject to criticism, it raises profound questions about the nature of the universe, the emergence of consciousness, and the limits of scientific explanation. Whether viewed as a speculative thought experiment or a serious challenge to conventional cosmological models, the Boltzmann Brain hypothesis serves as a reminder of the mysteries that still surround our understanding of reality and the cosmos.

Chapter 12: Dark Matter/Dark Energy

Dark matter and dark energy are two of the most mysterious and enigmatic components of the universe, comprising the vast majority of its mass-energy content. Despite their pervasive influence on the cosmos, they remain elusive and poorly understood. In this comprehensive exploration, we will delve into the historical context, observational evidence, theoretical frameworks, ongoing research efforts, and potential implications of dark matter and dark energy.

Historical Context:

The concept of dark matter can be traced back to the early 20th century, with the work of Swiss astronomer Fritz Zwicky in the 1930s. Zwicky studied the motions of galaxies within galaxy clusters and found that their observed velocities were much higher than what could be accounted for based on the visible matter alone. He hypothesized the existence of "dunkle Materie" or "dark matter" to explain this gravitational anomaly.

Dark energy, on the other hand, emerged as a concept much later, in the late 20th century. The discovery of the accelerating expansion of the universe in the late 1990s, based on observations of distant supernovae, provided the first evidence for the existence of a mysterious force driving the universe apart. This force was dubbed "dark energy" to distinguish it from ordinary matter and energy.

Observational Evidence:

1. **Galactic Rotation Curves**: Observations of the rotational velocities of galaxies, particularly spiral galaxies, have revealed that stars and gas in the outer regions of galaxies orbit at much higher speeds than expected based on the visible mass

alone. This discrepancy suggests the presence of unseen, or dark, matter exerting a gravitational influence.

2. **Galaxy Cluster Dynamics**: Studies of galaxy clusters, vast collections of galaxies bound together by gravity, have provided further evidence for dark matter. Measurements of the motions of galaxies within clusters, as well as the bending of light due to gravitational lensing, indicate the presence of unseen mass that far exceeds the mass of the visible galaxies.

3. **Cosmic Microwave Background**: The cosmic microwave background (CMB), relic radiation from the early universe, provides crucial information about its composition and evolution. Detailed measurements of the CMB, such as those obtained by the Planck satellite, have helped constrain the total amount of matter and energy in the universe, including dark matter and dark energy.

4. **Large-Scale Structure**: The distribution of galaxies and galaxy clusters across the universe, known as the large-scale structure, is influenced by the gravitational pull of dark matter. By mapping out the distribution of galaxies and measuring their clustering patterns, astronomers can infer the presence and distribution of dark matter.

Theoretical Frameworks:

1. **Cold Dark Matter**: The leading theoretical framework for dark matter is the cold dark matter (CDM) model, which posits that dark matter consists of slow-moving, non-interacting particles that formed shortly after the Big Bang. Candidates for cold dark matter include weakly interacting massive particles (WIMPs) and axions.

2. **Lambda Cold Dark Matter**: The Lambda Cold Dark Matter (ΛCDM) model extends the CDM model by incorporating dark energy, which is postulated to be responsible for the

observed accelerating expansion of the universe. In the ΛCDM model, dark energy accounts for about 68% of the total energy density of the universe, with dark matter comprising around 27%.

Ongoing Research Efforts:

1. **Particle Physics Experiments**: Scientists are conducting experiments in particle physics laboratories around the world to search for evidence of dark matter particles. These experiments involve detectors designed to detect the rare interactions between dark matter particles and ordinary matter.

2. **Astrophysical Observations**: Observatories such as the Hubble Space Telescope, the Atacama Large Millimeter/submillimeter Array (ALMA), and the upcoming James Webb Space Telescope (JWST) continue to study the properties of dark matter and its effects on the universe. These observations provide valuable insights into the distribution, composition, and behavior of dark matter.

3. **Cosmological Simulations**: Cosmologists use sophisticated computer simulations to model the formation and evolution of cosmic structures, such as galaxies and galaxy clusters, within the framework of dark matter and dark energy. These simulations help test theoretical models and compare their predictions with observational data.

4. **Gravitational Wave Astronomy**: The emerging field of gravitational wave astronomy, which studies the ripples in spacetime produced by cataclysmic events such as black hole mergers and neutron star collisions, offers a new way to probe the properties of dark matter and dark energy. Gravitational waves provide complementary information to electromagnetic observations, allowing scientists to study the

distribution of matter and energy in the universe.

Potential Implications:

1. **Fundamental Physics**: Understanding the nature of dark matter and dark energy could lead to breakthroughs in fundamental physics, such as the unification of gravity with the other fundamental forces and the development of a complete theory of quantum gravity.

2. **Cosmological Evolution**: Dark matter and dark energy play crucial roles in shaping the large-scale structure and evolution of the universe. A better understanding of these components could shed light on the fate of the universe, its ultimate destiny, and the possibilities for life elsewhere in the cosmos.

3. **Particle Physics**: Discovering the identity of dark matter particles would have profound implications for particle physics, potentially revealing new physics beyond the Standard Model and providing insights into the fundamental nature of matter and energy.

4. **Astrophysical Phenomena**: Dark matter and dark energy influence a wide range of astrophysical phenomena, from the formation of galaxies and galaxy clusters to the expansion history of the universe. Understanding their properties is essential for interpreting astronomical observations and testing cosmological models.

Conclusion:

Dark matter and dark energy represent two of the greatest mysteries in modern astrophysics and cosmology. Despite decades of research and observation, their true nature remains elusive, posing fundamental questions about the nature of the universe and our place within it. Continued efforts to study dark matter and dark energy, through both theoretical modeling and observational exploration, hold the promise of unlocking the secrets of the cosmos and deepening our understanding of the fundamental laws that govern the universe.

Chapter 13: Quantum Consciousness

Quantum consciousness is a theoretical framework that proposes the principles of quantum mechanics play a crucial role in the functioning of the human mind and consciousness. This idea, while controversial and still on the fringe of mainstream science, has intrigued scientists, philosophers, and thinkers for decades. The concept seeks to bridge the gap between the macroscopic world of classical physics, which governs everyday objects and phenomena, and the microscopic world of quantum physics, which governs the behavior of particles at the atomic and subatomic levels.

Historical Context

The concept of quantum consciousness has roots in the early 20th century, coinciding with the development of quantum mechanics. Quantum mechanics revolutionized our understanding of the physical world, introducing concepts such as wave-particle duality, superposition, and entanglement. These principles seemed to defy common sense and classical intuition, leading to philosophical debates about the nature of reality, observation, and measurement.

The idea that consciousness might have a quantum basis emerged in part due to the peculiarities of quantum mechanics. Notable physicists like Niels Bohr, Werner Heisenberg, and Erwin Schrödinger engaged in debates about the implications of quantum mechanics for reality and observation. Schrödinger's famous thought experiment, "Schrödinger's cat," exemplified the paradoxes inherent in quantum mechanics and the problem of measurement.

Key Concepts in Quantum Mechanics

1. **Wave-Particle Duality**: Particles such as electrons exhibit

both wave-like and particle-like properties. This duality is a cornerstone of quantum mechanics and suggests that the fundamental nature of particles is not as straightforward as previously thought.

2. **Superposition**: Quantum particles can exist in multiple states simultaneously until they are observed or measured. This principle is famously illustrated by Schrödinger's cat, which is simultaneously alive and dead until observed.

3. **Entanglement**: Particles can become entangled, meaning the state of one particle is directly related to the state of another, no matter the distance between them. This phenomenon implies a form of instantaneous communication that defies classical understanding of space and time.

4. **Quantum Tunneling**: Particles can pass through potential barriers that, according to classical physics, should be insurmountable. This phenomenon demonstrates the probabilistic nature of quantum mechanics.

Theoretical Proposals of Quantum Consciousness

Several theories propose mechanisms by which quantum mechanics might underpin consciousness. These theories often involve the interplay of quantum processes with the neurological activities of the brain. Key theories include:

1. **Penrose-Hameroff Orchestrated Objective Reduction (Orch-OR)**:
    - Proposed by physicist Sir Roger Penrose and anesthesiologist Stuart Hameroff.
    - Suggests that consciousness arises from quantum computations within microtubules, which are structural components of neurons.
    - According to Orch-OR, microtubules can exist in

quantum superpositions, and conscious experience results from the collapse of these superpositions via a process called "objective reduction."

- ○ Penrose and Hameroff argue that objective reduction is a fundamental process that occurs at the Planck scale, a level of physical reality where quantum gravity effects become significant.

2. **Quantum Brain Dynamics (QBD):**
   - ○ Suggests that the brain's neural processes can be understood as a form of quantum field theory.
   - ○ Proposes that coherent quantum states, similar to those observed in superconductors, could exist in the brain.
   - ○ These quantum states could be responsible for the integration of information across different regions of the brain, leading to coherent conscious experience.

3. **Holonomic Brain Theory:**
   - ○ Proposed by neuroscientist Karl Pribram and physicist David Bohm.
   - ○ Suggests that the brain operates like a hologram, processing information through wave interference patterns.
   - ○ Pribram and Bohm's theory is inspired by the holographic principle in physics, which posits that information about a three-dimensional space can be encoded on a two-dimensional surface.
   - ○ The theory implies that consciousness emerges from the brain's ability to process and integrate information in a holistic, non-local manner.

Experimental Evidence and Challenges

Despite the intriguing nature of these theories, empirical evidence for quantum consciousness is sparse and remains a significant challenge. Some key experimental approaches and findings include:

1. **Quantum Coherence in Biological Systems**:
    - Researchers have found evidence of quantum coherence and entanglement in biological systems, such as photosynthetic processes in plants and avian navigation.
    - These findings suggest that quantum effects can play a role in biological processes, lending some plausibility to the idea that similar effects could occur in the brain.
2. **Microtubules and Quantum Effects**:
    - Some studies have explored the potential for quantum effects in microtubules, the structures proposed by the Orch-OR theory.
    - However, the evidence is inconclusive, and many neuroscientists remain skeptical about the feasibility of sustained quantum coherence in the warm, wet, and noisy environment of the brain.
3. **Critiques of Quantum Mind Theories**:
    - Critics argue that quantum theories of consciousness often lack clear, testable predictions and are difficult to falsify.
    - The brain's macroscopic nature, with billions of neurons and synapses, seems incompatible with the delicate and highly sensitive nature of quantum states.
    - Many neuroscientists and physicists favor classical explanations for brain function and consciousness, viewing quantum theories as speculative and lacking

empirical support.

Philosophical Implications

The quantum consciousness hypothesis has profound philosophical implications, touching on the nature of reality, the mind-body problem, and the possibility of free will. Key philosophical considerations include:

1. **Nature of Reality**:
    ◦ Quantum mechanics challenges classical notions of objective reality, suggesting that observation plays a fundamental role in determining the state of a system.
    ◦ If consciousness is linked to quantum processes, it raises questions about the role of the observer in the universe and the nature of reality itself.
2. **Mind-Body Problem**:
    ◦ The mind-body problem concerns the relationship between the physical brain and subjective experience.
    ◦ Quantum consciousness theories propose that the mind cannot be fully explained by classical physical processes alone, potentially offering a new perspective on the nature of consciousness.
3. **Free Will**:
    ◦ Quantum mechanics introduces elements of indeterminacy and probability, which some interpret as allowing for a form of free will.
    ◦ If consciousness is rooted in quantum processes, it could imply that human decision-making involves elements of fundamental randomness, potentially reconciling free will with physical laws.

Implications for Neuroscience and AI

The exploration of quantum consciousness could have significant implications for neuroscience and artificial intelligence (AI). Some potential impacts include:

1. **Neuroscience**:
    - Understanding the role of quantum processes in the brain could lead to new insights into the mechanisms of consciousness and cognitive function.
    - It could also have implications for understanding and treating neurological disorders, providing new avenues for research and therapy.
2. **Artificial Intelligence**:
    - If consciousness arises from quantum processes, replicating these processes in AI systems could be necessary for achieving true artificial general intelligence (AGI).
    - Exploring quantum-based models of computation and information processing could inspire new approaches to developing intelligent machines.

Future Directions and Research

The field of quantum consciousness is still in its infancy, and many questions remain unanswered. Future research directions include:

1. **Empirical Studies**:
    - Developing experimental techniques to test for quantum effects in the brain and nervous system.
    - Investigating the potential for quantum coherence and entanglement in neural processes.
2. **Theoretical Developments**:

- Refining existing theories of quantum consciousness to make testable predictions and align with empirical data.
- Exploring new models and frameworks that integrate quantum mechanics with cognitive science and neuroscience.

3. **Interdisciplinary Collaboration**:
    - Fostering collaboration between physicists, neuroscientists, and philosophers to address the complex questions posed by quantum consciousness.
    - Encouraging dialogue between proponents and critics to advance the scientific understanding of consciousness.

Conclusion

Quantum consciousness represents a bold and intriguing hypothesis that seeks to explain the nature of consciousness through the principles of quantum mechanics. While it remains controversial and speculative, it has the potential to revolutionize our understanding of the mind and the fundamental nature of reality. Continued exploration and research in this field could lead to groundbreaking discoveries, challenging our perceptions and expanding the boundaries of human knowledge. As we delve deeper into the mysteries of consciousness and quantum mechanics, we may uncover new insights that reshape our understanding of ourselves and the universe.

Chapter 14: Dyson Sphere

The concept of a Dyson Sphere, an advanced megastructure that encompasses a star to capture a significant portion of its energy output, is one of the most intriguing and ambitious ideas in the realm of theoretical astrophysics and futurism. Named after the physicist and mathematician Freeman Dyson, who popularized the idea in a 1960 paper, the Dyson Sphere represents a hypothetical solution to the energy needs of a highly advanced civilization. This detailed exploration delves into the history, design concepts, feasibility, implications, and potential detection methods of a Dyson Sphere.

Historical Context and Origin

Freeman Dyson introduced the concept of what we now call a Dyson Sphere in his seminal 1960 paper, "Search for Artificial Stellar Sources of Infrared Radiation." Dyson speculated that a technologically advanced civilization, significantly more advanced than humanity, would eventually require energy far exceeding that available on its home planet. To meet these demands, such a civilization might build a massive structure around its star to capture its energy output.

Dyson's idea drew from earlier science fiction literature. Olaf Stapledon's 1937 novel "Star Maker" described similar constructs, and later science fiction works, such as Larry Niven's 1970 novel "Ringworld," expanded on the idea by presenting variations like the "Ringworld," a circular band structure around a star.

Concept and Design

The basic idea of a Dyson Sphere is to capture as much energy as possible from a star. There are several proposed designs for achieving this, ranging from solid shells to swarms of solar-collecting satellites. Here are the primary concepts:

1. **Dyson Swarm**:
   - The most plausible and commonly discussed form of a Dyson Sphere.
   - Consists of a large number of solar power satellites orbiting the star in a dense formation.
   - Each satellite would collect solar energy and transmit it to a central location for use.
   - This design avoids many of the structural and material challenges of a solid shell.
2. **Dyson Shell**:
   - A solid or nearly solid shell encircling the star at a distance where it would capture nearly all of the star's energy output.
   - The shell would have to be extremely large, with a radius of around 1 AU (the distance from the Earth to the Sun).
   - Structural integrity is a significant issue, as the shell would need to withstand immense forces.
3. **Dyson Bubble**:
   - A variant of the Dyson Swarm, consisting of solar sails rather than rigid satellites.
   - These solar sails would balance the star's radiation pressure against gravitational attraction to stay in position.
   - This design is less material-intensive but still captures substantial energy.
4. **Dyson Net**:
   - A network of interconnected platforms or cables forming a loose, web-like structure around the star.
   - Combines elements of swarms and shells, aiming to maximize energy collection while minimizing structural challenges.

Feasibility and Challenges

Constructing a Dyson Sphere poses immense technical and logistical challenges. Some of the key issues include:

1. **Materials and Engineering**:
   - The sheer scale of the structure requires materials with exceptional strength-to-weight ratios.
   - Current materials, even advanced composites and metals, may not suffice for the demands of a Dyson Shell.
   - Nanotechnology and molecular engineering could potentially provide solutions, but these are speculative.

2. **Energy Requirements**:
   - Building a Dyson Sphere would itself require vast amounts of energy.
   - Harvesting this energy would likely need the civilization to already have advanced solar or fusion power capabilities.

3. **Heat Dissipation**:
   - Capturing a star's energy means dealing with the resultant heat.
   - Effective methods for radiating or using this heat must be developed to prevent overheating and structural damage.

4. **Orbital Mechanics**:
   - Maintaining the stability of a Dyson Swarm or Bubble involves complex orbital mechanics to avoid collisions and ensure consistent energy capture.
   - The gravitational interactions among satellites and the star's gravitational field must be carefully managed.

5. **Economic and Resource Considerations**:
    - The resources required for a Dyson Sphere are beyond current human capacity.
    - Mining entire planets or asteroid belts might be necessary to gather sufficient materials.

Implications for Civilizations

Building a Dyson Sphere would place a civilization at the Kardashev Scale's Type II level, indicating mastery over planetary and stellar energy resources. This advancement has profound implications:

1. **Energy Abundance**:
    - A Dyson Sphere would provide near-limitless energy, supporting advanced technological, industrial, and potentially biological advancements.
    - This energy could power massive computational devices, space travel, terraforming, and other grand-scale projects.

2. **Technological Mastery**:
    - The construction and maintenance of such a structure imply a deep understanding of physics, materials science, and engineering.
    - Such a civilization would likely possess advanced AI, robotics, and possibly post-scarcity economic systems.

3. **Interstellar Influence**:
    - A Type II civilization with a Dyson Sphere would have significant influence over its stellar neighborhood.
    - It could potentially undertake large-scale interstellar colonization, spreading its presence across multiple star systems.

4. **Sociocultural Evolution**:
    ○ The societal and cultural implications are profound, as the civilization would undergo transformations to support and manage such vast projects.
    ○ This might include shifts in social structures, governance, and philosophical outlooks.

Potential for Detection

Detecting Dyson Spheres is an exciting prospect for astronomers searching for extraterrestrial intelligence (SETI). Potential methods include:

1. **Infrared Signatures**:
    ○ A Dyson Sphere would reradiate absorbed starlight as waste heat, primarily in the infrared spectrum.
    ○ Searching for unusual infrared emissions from stars that lack corresponding visible light signatures is a key strategy.
2. **Occlusion of Stars**:
    ○ Partial Dyson Spheres or swarms could cause observable dimming or fluctuations in a star's brightness.
    ○ Transit photometry, similar to methods used to detect exoplanets, could reveal these patterns.
3. **Spectral Anomalies**:
    ○ The material and construction of a Dyson Sphere might produce unique spectral signatures.
    ○ Analyzing star spectra for unusual features can provide clues.
4. **Radio Signals**:
    ○ Advanced civilizations might use powerful radio signals for communication.

- SETI initiatives focus on detecting these signals, which could originate from Dyson Sphere-building civilizations.

Case Studies and Hypotheses

There have been several intriguing cases where potential Dyson Sphere signatures were considered:

1. **Tabby's Star (KIC 8462852):**
   - Exhibited unusual dimming patterns, sparking speculation about the possibility of a Dyson Sphere or swarm.
   - Further studies suggested natural explanations, such as dust clouds or cometary activity, but the mystery remains unresolved.
2. **Search for Infrared Excess:**
   - Surveys like the Wide-field Infrared Survey Explorer (WISE) search for stars with excess infrared emissions.
   - While no definitive Dyson Spheres have been identified, these searches continue to refine techniques and broaden our understanding.

Philosophical and Ethical Considerations

The idea of constructing a Dyson Sphere raises profound philosophical and ethical questions:

1. **Existential Purpose:**
   - The pursuit of such grand projects reflects a civilization's existential drive for survival, expansion, and exploration.
   - It raises questions about the ultimate purpose and

destiny of intelligent life.

2. **Environmental Impact**:
   - The transformation of entire star systems has significant environmental implications.
   - Ethical considerations about the disruption of potential ecosystems and the long-term sustainability of such projects must be addressed.
3. **Civilizational Responsibility**:
   - A civilization capable of building a Dyson Sphere holds immense power.
   - The responsibility of managing this power without causing harm to itself or other potential civilizations is paramount.

Future Prospects and Research Directions

The concept of the Dyson Sphere continues to inspire both scientific inquiry and speculative fiction. Future research directions include:

1. **Advancements in Material Science**:
   - Developing new materials with the necessary properties for constructing megastructures.
   - Nanotechnology and molecular engineering hold promise for breakthroughs.
2. **Astrophysical Observations**:
   - Expanding infrared and radio surveys to detect potential Dyson Spheres.
   - Improving data analysis techniques to identify subtle anomalies in stellar behavior.
3. **Theoretical Models**:
   - Refining theoretical models of Dyson Sphere construction, stability, and energy management.
   - Exploring the implications for astrobiology and the

search for extraterrestrial intelligence.

4. **Interdisciplinary Collaboration**:
    - Encouraging collaboration between physicists, engineers, astronomers, and ethicists to address the multifaceted challenges of Dyson Sphere research.
    - Integrating insights from science fiction to stimulate innovative thinking and public engagement.

Conclusion

The concept of a Dyson Sphere remains one of the most captivating and ambitious ideas in science and futurism. It challenges our understanding of technological progress, the potential of intelligent life, and the ultimate destiny of civilizations. While the practical realization of a Dyson Sphere lies far in the future, exploring this idea pushes the boundaries of our imagination and scientific inquiry. As we continue to search for evidence of extraterrestrial intelligence and expand our technological capabilities, the dream of harnessing the full power of a star might one day become a reality, heralding a new era of cosmic exploration and discovery.

Chapter 15: Biocentrism

Biocentrism is a philosophical and scientific theory that places life and consciousness at the center of our understanding of the universe. Proposed by Robert Lanza, a prominent American scientist, this paradigm challenges the traditional materialistic view that the universe exists independently of observation and life. Instead, biocentrism posits that life and consciousness are fundamental to the fabric of reality, suggesting that the universe is shaped by the perception of living beings.

Historical Context and Development

Biocentrism emerged as a response to the limitations and paradoxes inherent in traditional scientific and philosophical frameworks. The materialistic view, which has dominated science since the Enlightenment, posits that the universe is a vast, mechanistic entity governed by immutable physical laws. In this view, life and consciousness are byproducts of complex arrangements of matter and energy.

Robert Lanza introduced biocentrism in his 2007 book "Biocentrism: How Life and Consciousness Are the Keys to Understanding the True Nature of the Universe." He argues that the traditional view fails to account for the role of consciousness in shaping our understanding of reality. Drawing on principles from quantum mechanics, neuroscience, and biology, Lanza proposes that life and consciousness are not accidental byproducts but rather integral to the structure of the universe.

Core Principles of Biocentrism

Biocentrism is grounded in several core principles that redefine our understanding of space, time, and the nature of reality. These principles include:

1. **The Principle of Consciousness**:
     - Consciousness is fundamental to the universe and precedes the material world.
     - Reality cannot exist without the presence of conscious observers to perceive it.
2. **The Role of Observation**:
     - The act of observation shapes the very fabric of reality.
     - Quantum mechanics supports this principle, where particles exist in a superposition of states until observed.
3. **Space and Time as Constructs**:
     - Space and time are not absolute entities but constructs of the conscious mind.
     - They do not exist independently of life and perception but are tools created by consciousness to understand the world.
4. **The Universe as Biocentric**:
     - The universe is fine-tuned to support life and consciousness.
     - The constants and laws of nature are precisely calibrated to allow the existence of observers.

Quantum Mechanics and Biocentrism

One of the most compelling aspects of biocentrism is its alignment with the principles of quantum mechanics, which often defy classical intuitions about reality. Key quantum phenomena that support biocentrism include:

1. **Wave-Particle Duality**:
    ◦ Particles such as electrons and photons exhibit both wave-like and particle-like properties depending on how they are observed.
    ◦ This duality suggests that the nature of reality is influenced by the observer's choices.
2. **Quantum Superposition**:
    ◦ Quantum particles can exist in multiple states simultaneously until measured.
    ◦ The famous double-slit experiment demonstrates that particles behave differently when observed, implying that observation affects reality.
3. **Quantum Entanglement**:
    ◦ Entangled particles exhibit correlations that cannot be explained by classical physics, regardless of the distance separating them.
    ◦ This phenomenon implies a non-local connection that defies classical notions of space and time.
4. **The Observer Effect**:
    ◦ The act of measurement or observation collapses a particle's wave function into a definite state.
    ◦ This principle underscores the central role of the observer in determining the outcome of quantum events.

Neuroscience and Consciousness

Biocentrism also draws on insights from neuroscience, emphasizing the importance of consciousness in shaping our perception of reality. Key points include:

1. **Perception and Reality**:
    ◦ Our sensory systems and brain processes construct

our experience of the world.

- Colors, sounds, and other sensory experiences are not inherent properties of objects but are created by the brain in response to stimuli.

2. **The Binding Problem**:
    - The binding problem in neuroscience concerns how disparate sensory inputs are integrated into a cohesive conscious experience.
    - Biocentrism suggests that this integration is fundamental to the nature of reality, emphasizing the active role of consciousness in shaping experience.

3. **Consciousness as a Fundamental Entity**:
    - Rather than being an emergent property of neural activity, consciousness is proposed to be a fundamental aspect of the universe.
    - This view challenges reductionist approaches that attempt to explain consciousness solely through physical processes.

Philosophical Implications

Biocentrism has profound philosophical implications, challenging long-held assumptions about the nature of reality, life, and consciousness. Key philosophical considerations include:

1. **The Nature of Reality**:
    - If consciousness is fundamental, reality itself is subjective and observer-dependent.
    - This perspective aligns with idealism, a philosophical view that reality is mentally constructed.

2. **Epistemology**:
    - Biocentrism redefines how we acquire knowledge, emphasizing the central role of conscious experience.

- ○ It challenges the objective, observer-independent perspective of traditional science.

3. **Anthropic Principle**:
   - ○ The anthropic principle states that the universe's physical laws and constants are fine-tuned to allow the existence of observers.
   - ○ Biocentrism extends this idea, suggesting that the universe is fundamentally shaped by the presence of life and consciousness.

4. **Existential and Ethical Implications**:
   - ○ Recognizing life and consciousness as central to the universe can influence our ethical and existential perspectives.
   - ○ It promotes a view of humans and other conscious beings as integral to the cosmos, potentially fostering a deeper sense of connection and responsibility.

Criticisms and Counterarguments

Biocentrism, like any groundbreaking theory, faces several criticisms and counterarguments. Key points of contention include:

1. **Scientific Testability**:
   - ○ Critics argue that biocentrism lacks empirical testability and falsifiability, making it more of a philosophical stance than a scientific theory.
   - ○ Proponents respond by highlighting the theory's consistency with quantum mechanics and the need for new paradigms to address the mysteries of consciousness.

2. **Reductionism**:
   - ○ Traditional scientific approaches favor reductionism, explaining complex phenomena in terms of simpler,

underlying processes.

- ○ Biocentrism challenges this approach, suggesting that consciousness cannot be fully understood through reductionist methods.

3. **Alternative Theories of Consciousness**:
    - ○ Other theories, such as emergentism and physicalism, propose that consciousness arises from complex interactions of physical processes.
    - ○ Biocentrism's claim of consciousness as fundamental is seen by some as an unnecessary departure from these well-established frameworks.

4. **Philosophical Challenges**:
    - ○ Biocentrism's idealistic tendencies clash with philosophical materialism, which posits that matter and energy constitute the primary reality.
    - ○ Critics argue that biocentrism risks falling into solipsism, the view that only one's mind is sure to exist.

Implications for Science and Future Research

Despite the criticisms, biocentrism offers intriguing avenues for future research and has potential implications for various scientific fields. Key areas of interest include:

1. **Quantum Biology**:
    - ○ Exploring quantum effects in biological systems and their potential role in consciousness.
    - ○ Investigating whether quantum processes are integral to neural activity and perception.

2. **Neuroscience and Cognitive Science**:
    - ○ Developing new models of consciousness that integrate biocentric principles.

- Studying the relationship between brain activity, subjective experience, and the construction of reality.

3. **Cosmology**:
   - Reevaluating cosmological models to account for the role of observers in shaping the universe.
   - Investigating the anthropic principle and the fine-tuning of physical constants in a biocentric context.

4. **Interdisciplinary Collaboration**:
   - Fostering collaboration between physicists, biologists, neuroscientists, and philosophers to explore the implications of biocentrism.
   - Integrating insights from various disciplines to develop a holistic understanding of consciousness and reality.

Conclusion

Biocentrism represents a radical shift in our understanding of the universe, placing life and consciousness at the core of reality. By challenging the traditional materialistic paradigm, biocentrism invites us to reconsider the nature of existence, the role of observers, and the fundamental principles that govern the cosmos. While the theory faces significant challenges and criticisms, it also opens up new avenues for exploration and offers a fresh perspective on some of the most profound questions in science and philosophy. As research continues and our understanding of quantum mechanics, neuroscience, and cosmology deepens, biocentrism may provide valuable insights into the nature of consciousness and the universe.

Chapter 16: Anthropic Principle

The Anthropic Principle is a philosophical consideration that the universe's physical laws and constants are fine-tuned in a way that allows for the existence of life, particularly intelligent life capable of observing and reflecting on the universe itself. The principle is derived from the Greek word "anthropos," meaning "human," and it suggests that the universe must be compatible with the conscious and sapient life that observes it. This principle has significant implications for cosmology, physics, and philosophy, leading to various interpretations and debates. This detailed exploration delves into the history, types, implications, criticisms, and ongoing research related to the Anthropic Principle.

Historical Context and Development

The Anthropic Principle emerged from discussions in cosmology and physics about the apparent fine-tuning of the universe. The term was first introduced by Brandon Carter, a theoretical physicist, in 1973 during a conference celebrating the 500th birthday of Copernicus. Carter proposed that the universe's fundamental constants are constrained by the necessity to allow the existence of observers within it.

The principle gained further attention through the work of physicists and cosmologists like John D. Barrow and Frank J. Tipler, who elaborated on the concept in their 1986 book "The Anthropic Cosmological Principle." They categorized the principle into different forms and discussed its implications for understanding the universe.

Types of the Anthropic Principle

The Anthropic Principle is broadly categorized into several forms, each with its nuances and implications:

1. **Weak Anthropic Principle (WAP):**
    ◦ States that the observed values of physical and cosmological parameters are restricted by the requirement that they allow for the existence of observers.
    ◦ In essence, we observe the universe to be compatible with our existence because if it were not, we would not be here to observe it.
    ◦ Often used to explain the apparent fine-tuning without invoking any deeper significance.
2. **Strong Anthropic Principle (SAP):**
    ◦ Suggests that the universe must have properties that inevitably lead to the development of life at some stage.
    ◦ Implies that the existence of life is not merely a coincidence but a fundamental aspect of the universe's structure.
3. **Final Anthropic Principle (FAP):**
    ◦ Proposed by Barrow and Tipler, this principle asserts that intelligent life must come into existence in the universe and, once it comes into existence, will never die out.
    ◦ It suggests a form of cosmic destiny where intelligent life is an essential component of the universe's future.
4. **Participatory Anthropic Principle (PAP):**
    ◦ Proposed by physicist John Archibald Wheeler, this principle suggests that observers are necessary to bring the universe into being.
    ◦ It aligns with certain interpretations of quantum mechanics, where observation plays a crucial role in the manifestation of reality.
5. **Anthropic Selection Principle (ASP):**

- Focuses on the idea that our universe is just one of many possible universes in a multiverse, and we find ourselves in one where conditions are right for life.
- This principle is often invoked in the context of the multiverse hypothesis.

Implications of the Anthropic Principle

The Anthropic Principle has far-reaching implications across various domains, from cosmology to philosophy and theology:

1. **Cosmology and Fine-Tuning**:
   - The principle addresses the fine-tuning problem, where the physical constants and initial conditions of the universe appear to be precisely calibrated to allow life.
   - For instance, the strength of the gravitational force, the electromagnetic force, and the rate of expansion of the universe all fall within narrow ranges that permit the formation of stars, planets, and ultimately life.
2. **Philosophy and Metaphysics**:
   - The Anthropic Principle raises questions about the nature of reality and our place in the universe.
   - It challenges the notion of an objective, observer-independent universe by suggesting that our observations are inextricably linked to the existence of life.
3. **Theology and Teleology**:
   - Some interpretations of the Anthropic Principle align with theological and teleological views, suggesting that the universe is designed or purposefully configured to support life.

- This perspective resonates with arguments for intelligent design and the idea of a purposeful creation.

4. **Multiverse Hypothesis**:
   - The Anthropic Principle is often used to support the multiverse hypothesis, which posits the existence of many universes with varying physical laws and constants.
   - In this context, our universe is just one of many, and we observe it to be life-permitting because we could not exist in universes where the conditions are not suitable for life.

Criticisms and Counterarguments

The Anthropic Principle, despite its intriguing implications, has faced several criticisms and counterarguments:

1. **Tautology and Triviality**:
   - Critics argue that the Weak Anthropic Principle is tautological and trivial, as it merely states that we observe the universe to be life-permitting because we are here to observe it.
   - This form of the principle does not provide any explanatory power or predictive capability.

2. **Lack of Testability**:
   - The principle is criticized for being untestable and unfalsifiable, as it does not make specific predictions that can be empirically verified.
   - This lack of scientific rigor makes it more of a philosophical statement than a scientific theory.

3. **Anthropic Bias**:
   - Some critics argue that the principle introduces an

anthropic bias, prioritizing human existence as a fundamental aspect of the universe.

- This bias can lead to the anthropic fallacy, where human-centered perspectives are incorrectly assumed to be central to the nature of the cosmos.

4. **Alternative Explanations**:
   - There are alternative explanations for the fine-tuning of the universe that do not rely on the Anthropic Principle.
   - For example, some theories suggest that the physical constants are not as fine-tuned as they appear, or that there are natural mechanisms that lead to the observed values.

Ongoing Research and Future Directions

Despite the criticisms, the Anthropic Principle continues to be a topic of interest and debate in scientific and philosophical circles. Ongoing research and future directions include:

1. **Cosmological Observations**:
   - Advances in observational cosmology, such as the study of the cosmic microwave background radiation and the distribution of galaxies, can provide insights into the fundamental constants and the conditions of the early universe.
   - These observations can help refine our understanding of the fine-tuning problem and the applicability of the Anthropic Principle.

2. **Theoretical Physics**:
   - Developments in theoretical physics, particularly in the areas of quantum gravity and string theory, may offer new perspectives on the fundamental nature of

the universe and the role of observers.

- The exploration of higher-dimensional spaces and the potential for multiple universes in string theory can provide a framework for understanding the Anthropic Principle.

3. **Multiverse Hypothesis**:
    - The investigation of the multiverse hypothesis continues to be a fertile area of research, with implications for the Anthropic Principle.
    - Theoretical models and simulations of multiverse scenarios can shed light on the distribution of life-permitting universes and the statistical likelihood of our own universe.

4. **Philosophical Inquiry**:
    - Philosophers continue to explore the implications of the Anthropic Principle for our understanding of reality, knowledge, and existence.
    - The principle invites interdisciplinary dialogue between scientists, philosophers, and theologians, fostering a deeper exploration of the nature of the universe and our place within it.

Conclusion

The Anthropic Principle offers a compelling perspective on the universe, highlighting the intricate relationship between the cosmos and the observers within it. By suggesting that the universe is fine-tuned to support life, the principle challenges traditional views of an objective, observer-independent reality and raises profound questions about the nature of existence. While the principle faces significant criticisms and counterarguments, it continues to inspire scientific inquiry, philosophical reflection, and theological debate. As our understanding of the universe evolves, the Anthropic Principle remains a central and thought-provoking concept, inviting us to reconsider the fundamental nature of the cosmos and our place within it.

Chapter 17: Cosmic Inflation Theory

Cosmic inflation theory stands as one of the most influential and revolutionary concepts in modern cosmology. It proposes that the universe underwent an extremely rapid expansion phase in the very first moments after the Big Bang, exponentially increasing its size and smoothing out irregularities. This theory addresses various puzzles in cosmology, including the uniformity of the cosmic microwave background radiation, the large-scale structure of the universe, and the horizon problem. This comprehensive exploration delves into the origins, mechanisms, evidence, implications, and ongoing research related to cosmic inflation theory.

Origins and Development

Cosmic inflation theory originated from the work of physicist Alan Guth in the late 1970s. Guth was attempting to address the horizon problem, which refers to the apparent uniformity of the cosmic microwave background (CMB) radiation observed from opposite ends of the observable universe. According to conventional cosmology, regions of space that are now widely separated were never in causal contact, yet they exhibit remarkably similar properties.

Guth proposed that the universe underwent a brief period of exponential expansion, driven by a hypothetical field called the inflaton. This expansion, occurring within the first fraction of a second after the Big Bang, would have stretched the universe beyond the horizon distance, effectively resolving the horizon problem.

Mechanisms of Inflation

Cosmic inflation relies on the dynamics of a scalar field known as the inflaton. The key features of inflationary models include:

1. **Inflaton Potential**:
    - The inflaton field is characterized by a potential energy function that determines its behavior.
    - During the inflationary phase, the inflaton field undergoes a temporary state of high energy density, driving the rapid expansion of space.
2. **Exponential Expansion**:
    - The inflaton field experiences a period of slow roll, during which its potential energy dominates over its kinetic energy.
    - This leads to a phase of exponential expansion, where the scale factor of the universe increases at an accelerating rate.
3. **End of Inflation**:
    - Inflation ends when the inflaton field reaches a critical value or undergoes a phase transition, releasing its stored energy.
    - This energy is converted into matter and radiation, initiating the hot Big Bang phase of the universe's evolution.

Evidence for Cosmic Inflation

Despite its brief duration, cosmic inflation has left several observable imprints on the universe, providing indirect evidence for its occurrence:

1. **Flatness of the Universe**:
    - Inflationary models predict that the geometry of the universe should be flat, with Euclidean geometry on large scales.
    - Observations of the cosmic microwave background and large-scale structure support this prediction,

indicating that the universe is indeed flat to within a small margin of error.

2. **Homogeneity and Isotropy**:
   - Inflationary expansion would have smoothed out irregularities in the early universe, leading to its observed homogeneity and isotropy on large scales.
   - Anisotropies in the CMB, measured by satellites like the Cosmic Background Explorer (COBE) and the Wilkinson Microwave Anisotropy Probe (WMAP), match theoretical predictions based on inflationary models.

3. **Primordial Gravitational Waves**:
   - Inflation generates gravitational waves as a result of quantum fluctuations in spacetime during the rapid expansion phase.
   - These primordial gravitational waves imprint a characteristic polarization pattern on the CMB, known as B-mode polarization.
   - The detection of B-mode polarization by experiments like the BICEP/Keck and the Planck satellite provides indirect evidence for inflation.

Implications of Cosmic Inflation

Cosmic inflation has profound implications for our understanding of the universe's origins, structure, and evolution:

1. **Resolution of Cosmological Puzzles**:
   - Inflationary models provide elegant solutions to long-standing problems in cosmology, such as the horizon problem, flatness problem, and monopole problem.
   - By invoking a brief period of rapid expansion,

inflation can account for the observed properties of the universe without the need for ad hoc modifications.

2. **Origin of Cosmic Structure**:
   - Quantum fluctuations during inflation serve as seeds for the formation of structure in the universe.
   - Density variations imprinted during inflation give rise to the large-scale structure of galaxies, clusters, and cosmic voids observed today.

3. **Multiverse Hypothesis**:
   - Inflationary models naturally lead to the concept of a multiverse, where different regions of space undergo independent periods of inflation.
   - The multiverse hypothesis has implications for the anthropic principle and the fine-tuning of the universe's physical constants.

4. **Particle Physics and Grand Unified Theories**:
   - Inflation occurs at energies much higher than those accessible to particle accelerators, providing a unique probe of physics at the highest energy scales.
   - The dynamics of inflation are closely related to the properties of elementary particles and their interactions, offering insights into grand unified theories of particle physics.

Criticisms and Challenges

Despite its successes, cosmic inflation theory is not without its challenges and criticisms:

1. **Fine-Tuning of Inflaton Potential**:
   - Inflationary models require specific conditions for the inflaton potential to drive the necessary period of

exponential expansion.

- ◦ Critics argue that the fine-tuning of these parameters raises questions about the naturalness of inflation.

2. **Initial Conditions and Eternal Inflation**:
   - ◦ Explaining the initial conditions necessary for inflation to begin poses challenges, as inflation itself does not provide a mechanism for setting these conditions.
   - ◦ Some inflationary models lead to eternal inflation, where inflation never completely ends, raising questions about how to define the beginning of our universe.

3. **Testability and Falsifiability**:
   - ◦ While inflation makes specific predictions about the statistical properties of the CMB and large-scale structure, some aspects of inflationary models are difficult to test observationally.
   - ◦ Critics argue that inflation may be too flexible, allowing for a wide range of possible outcomes that can fit observational data.

4. **Alternative Cosmological Scenarios**:
   - ◦ There exist alternative cosmological scenarios that can account for the observed properties of the universe without invoking inflation.
   - ◦ Some modified gravity theories and bouncing cosmologies propose different mechanisms for generating the observed features of the universe.

Ongoing Research and Future Directions

Research on cosmic inflation continues to be a vibrant and active area of investigation, with ongoing efforts focused on:

1. **Precision Cosmology**:
    - Improving measurements of the cosmic microwave background and large-scale structure to test inflationary predictions with greater precision.
    - Future experiments like the Simons Observatory and the Cosmic Origins Explorer (CORE) satellite aim to map the CMB with unprecedented detail.
2. **Inflationary Model Building**:
    - Developing and refining inflationary models that can explain observational data while addressing theoretical challenges.
    - Exploring extensions of inflation, such as warm inflation and multifield inflation, to accommodate new observational constraints.
3. **Gravitational Wave Detection**:
    - Detecting primordial gravitational waves directly using experiments like the Laser Interferometer Gravitational-Wave Observatory (LIGO) and the European Space Agency's Laser Interferometer Space Antenna (LISA).
    - Direct detection of gravitational waves from inflation would provide smoking-gun evidence for the theory.
4. **Fundamental Physics Connections**:
    - Exploring the connections between inflation, particle physics, and fundamental theories such as string theory and quantum gravity.
    - Investigating how inflationary dynamics emerge from more fundamental theories of quantum gravity, such as loop quantum cosmology and string theory.
    - Exploring the implications of inflation for the nature of spacetime, quantum gravity effects, and the origin

of the universe's initial conditions.

Conclusion

Cosmic inflation theory represents a remarkable synthesis of theoretical physics and observational cosmology, providing a compelling framework for understanding the early universe's dynamics and evolution. By proposing a period of rapid expansion in the universe's infancy, inflationary models offer elegant solutions to longstanding cosmological puzzles and make precise predictions that can be tested with observational data.

Despite facing challenges and criticisms, cosmic inflation theory has garnered substantial empirical support from observations of the cosmic microwave background, large-scale structure, and gravitational waves. Ongoing research continues to refine our understanding of inflationary dynamics, probe its connections to fundamental physics, and explore its implications for the nature of reality.

As we delve deeper into the mysteries of the cosmos and push the boundaries of observational and theoretical cosmology, cosmic inflation remains a central pillar of our cosmological framework, offering profound insights into the origin, structure, and fate of the universe. With continued advancements in technology, theory, and observation, cosmic inflation theory promises to unlock even deeper secrets of the cosmos and further illuminate our cosmic origins.

Chapter 18: Fermi Bubble Hypothesis

The Fermi Bubble Hypothesis proposes that two enormous structures, known as the Fermi Bubbles, extend above and below the Milky Way's galactic plane. These bubbles, discovered in 2010 by the Fermi Gamma-ray Space Telescope, stretch for tens of thousands of light-years and are composed of high-energy gamma-ray emissions.

Discovery and Characteristics

The Fermi Bubbles were first detected in gamma-ray emissions by the Fermi Gamma-ray Space Telescope, operated by NASA. These bubbles are symmetrically positioned above and below the center of the Milky Way galaxy, extending perpendicular to the galactic plane.

The bubbles exhibit several key characteristics:

1. **Size and Scale**:
    - The Fermi Bubbles are vast structures, extending over 25,000 light-years from their center to the edges.
    - They span a significant portion of the Milky Way galaxy, with a diameter comparable to the size of the galaxy itself.
2. **Gamma-Ray Emissions**:
    - The Fermi Bubbles emit high-energy gamma rays, indicating the presence of energetic processes within them.
    - These gamma-ray emissions are primarily observed in the GeV (billion electron volts) energy range.
3. **Symmetry and Orientation**:
    - The bubbles are symmetrically positioned above and below the galactic center, aligned with the Milky Way's axis of rotation.

- Their orientation suggests a connection to processes occurring at the galactic center, such as star formation or black hole activity.

Proposed Origins and Mechanisms

Several hypotheses have been proposed to explain the formation and nature of the Fermi Bubbles:

1. **Galactic Winds**:
   - One leading explanation suggests that the bubbles are the result of galactic-scale winds driven by intense star formation and black hole activity at the Milky Way's center.
   - These winds, powered by energy released from massive stars and supermassive black holes, could blow gas and cosmic rays out of the galactic plane, creating the observed structures.

2. **Activity of the Milky Way's Supermassive Black Hole**:
   - The Milky Way harbors a supermassive black hole, known as Sagittarius A*, at its center.
   - Intense activity, such as accretion of matter and the ejection of jets, from Sagittarius A* could produce the energetic outflows responsible for forming the Fermi Bubbles.

3. **Cosmic Ray Acceleration**:
   - Another hypothesis posits that the Fermi Bubbles are the result of processes that accelerate cosmic rays to extreme energies.
   - Shock waves produced by supernova explosions or interactions between stellar winds and interstellar gas could accelerate cosmic rays and generate gamma-ray emissions.

4. **Historical Galactic Activity**:
   - The Fermi Bubbles may be remnants of past galactic activity, such as energetic outbursts from the Milky Way's central black hole or episodes of intense star formation.
   - These historical events could have left behind the observed structures as signatures of the galaxy's dynamic evolution.

Observational Evidence and Studies

Observations from various telescopes and instruments have provided valuable insights into the nature of the Fermi Bubbles:

1. **Gamma-Ray Surveys**:
   - The Fermi Gamma-ray Space Telescope has conducted extensive surveys of the gamma-ray sky, revealing the presence of the Fermi Bubbles and mapping their morphology and spectral characteristics.
2. **Multi-Wavelength Studies**:
   - Observations across different wavelengths, including X-ray, infrared, and radio, have helped astronomers characterize the Fermi Bubbles and understand their relationship to other components of the Milky Way.
3. **High-Energy Particle Detection**:
   - Instruments designed to detect high-energy particles, such as cosmic rays and neutrinos, have provided additional evidence for the presence of energetic processes associated with the Fermi Bubbles.
4. **Simulations and Modeling**:
   - Computational models and simulations have been used to simulate the formation and evolution of the

Fermi Bubbles, testing various hypotheses and comparing simulated results with observational data.

Implications and Significance

The discovery of the Fermi Bubbles has significant implications for our understanding of the Milky Way galaxy and galactic-scale phenomena:

1. **Galactic Dynamics**:
    - The presence of the Fermi Bubbles highlights the dynamic nature of the Milky Way galaxy and the importance of energetic processes in shaping its structure and evolution.
2. **Cosmic Ray Astronomy**:
    - Studying the Fermi Bubbles provides insights into the origins and acceleration mechanisms of cosmic rays, which are high-energy particles that pervade the universe.
3. **Feedback Mechanisms**:
    - The energetic outflows associated with the Fermi Bubbles may play a role in regulating star formation and galactic dynamics by exerting feedback on the interstellar medium.
4. **Galactic Archaeology**:
    - Understanding the formation and history of the Fermi Bubbles can shed light on past events and processes that have shaped the Milky Way over billions of years.

Future Research Directions

Ongoing and future research on the Fermi Bubbles aims to address key unanswered questions and further our understanding of these enigmatic structures:

1. **Origin and Formation**:
   - Researchers seek to determine the precise mechanisms responsible for creating the Fermi Bubbles, including the role of galactic winds, black hole activity, and cosmic ray acceleration.
2. **Evolution and Dynamics**:
   - Studying the temporal evolution of the Fermi Bubbles and their interactions with the surrounding interstellar medium can provide insights into their long-term dynamics and fate.
3. **Multi-Messenger Astronomy**:
   - Combining observations across multiple wavelengths and using complementary techniques, such as neutrino and gravitational wave detection, can provide a more comprehensive understanding of the Fermi Bubbles and their associated phenomena.
4. **Theoretical Modeling**:
   - Refining theoretical models and simulations to better match observational data and test competing hypotheses will be crucial for advancing our understanding of the Fermi Bubbles' origins and properties.

Conclusion

The Fermi Bubble Hypothesis represents a fascinating area of research in modern astrophysics, offering insights into the dynamic and energetic processes at play in the Milky Way galaxy. While significant progress has been made in understanding these enigmatic structures, many questions remain unanswered. Continued observational efforts, theoretical modeling, and interdisciplinary research will be essential for unraveling the mysteries of the Fermi Bubbles and their role in shaping the cosmos.

Chapter 19: The Big Crunch

The Big Crunch theory is a cosmological hypothesis proposing that the expansion of the universe, which began with the Big Bang, will eventually reverse, leading to a collapse of the universe back into a singularity. This concept stands in contrast to the more widely accepted theory of the universe's continued expansion, known as the Big Freeze or Heat Death scenario. In this detailed exploration, we will delve into the origins, implications, evidence, criticisms, and alternative scenarios related to the Big Crunch theory.

Origins and Development

The idea of a cyclic universe, where periods of expansion are followed by contraction, has roots in ancient cosmological theories. However, the modern concept of the Big Crunch emerged from the equations of general relativity and gained prominence in the mid-20th century.

One of the earliest proponents of the Big Crunch was Richard Tolman, who, in 1934, proposed a cyclic model of the universe where cycles of expansion and contraction occur indefinitely. However, it was not until the late 1960s and early 1970s that the Big Crunch gained widespread attention in the context of the expanding universe model.

Mechanism of the Big Crunch

The Big Crunch is driven by the gravitational attraction between all matter and energy in the universe. According to the theory, the expansion of the universe, which began with the Big Bang, gradually slows down due to the gravitational pull of matter and dark energy. Eventually, this deceleration leads to a reversal of the expansion, culminating in a collapse of the universe back into a singularity.

The process of the Big Crunch can be described in several stages:

1. **Expansion Phase**:
     ◦ The universe begins with the Big Bang, expanding rapidly and cooling over billions of years.
     ◦ During this phase, galaxies move away from each other as the fabric of space itself expands.
2. **Deceleration**:
     ◦ As the universe expands, the gravitational pull of matter and dark energy gradually slows down the expansion.
     ◦ Eventually, the rate of expansion becomes insufficient to overcome the gravitational attraction between galaxies.
3. **Reversal of Expansion**:
     ◦ Once the expansion slows down to a critical point, the gravitational attraction between galaxies causes the expansion to reverse.
     ◦ Galaxies begin to move towards each other, and the universe contracts.
4. **Collapse into Singularity**:
     ◦ The contraction continues until all matter and energy in the universe converge into a single point of infinite density and temperature, known as a singularity.
     ◦ This marks the end of the current cycle of the universe, with the potential for a new Big Bang and a new cycle of expansion.

Evidence for the Big Crunch

The evidence for the Big Crunch theory is limited, and observational data overwhelmingly supports the idea of the universe's continued expansion rather than a future collapse. However, several lines of evidence have been proposed in support of the Big Crunch:

1. **Critical Density**:
    - The fate of the universe depends on its density relative to a critical value known as the critical density.
    - If the actual density of the universe exceeds the critical density, it would eventually lead to a Big Crunch.
    - Observations of the cosmic microwave background radiation and large-scale structure suggest that the density of the universe may be close to the critical value.
2. **Closed Universe Geometry**:
    - The geometry of the universe can provide clues about its ultimate fate.
    - A closed universe, where space curves back on itself like the surface of a sphere, is consistent with the possibility of a Big Crunch.
    - Observational studies of the universe's geometry have yielded inconclusive results, with some indications of a closed universe and others suggesting flat or open geometries.
3. **Dark Energy Dynamics**:
    - The behavior of dark energy, a mysterious form of energy thought to drive the universe's accelerated expansion, could influence the fate of the universe.
    - If dark energy behaves differently than currently understood, it could potentially lead to a future contraction and a Big Crunch scenario.

Implications of the Big Crunch

The Big Crunch theory has significant implications for the fate of the universe and our understanding of cosmology:

1. **Cyclic Universe**:
   - If the Big Crunch were to occur, it would imply a cyclic model of the universe, where periods of expansion are followed by contraction and vice versa.
   - This cyclic nature of the universe has profound implications for our understanding of time, space, and cosmic evolution.
2. **Ultimate Destiny**:
   - The idea of a Big Crunch raises questions about the ultimate destiny of the universe and whether it will experience a finite or infinite existence.
   - Understanding the fate of the universe has implications for philosophical and existential inquiries into the nature of reality and existence.
3. **Cosmic Evolution**:
   - The occurrence of a Big Crunch would mark a dramatic phase transition in the evolution of the universe, from expansion to contraction.
   - Studying this transition can provide insights into the fundamental forces and dynamics that govern the universe on the largest scales.

Criticisms and Alternative Scenarios

Despite its conceptual appeal, the Big Crunch theory faces several criticisms and challenges:

1. **Observational Evidence**:
   - Observational data overwhelmingly supports the idea of the universe's continued expansion, with no definitive evidence for a future collapse.
   - Measurements of the cosmic microwave background, the Hubble constant, and the large-scale distribution

of galaxies all point to an expanding universe.

2. **Dark Energy Dominance**:
    ○ The discovery of dark energy and its observed effect of accelerating the universe's expansion has shifted the cosmological paradigm away from the Big Crunch scenario.
    ○ Dark energy's repulsive effect on cosmic scales makes a future collapse less likely, as it counteracts the gravitational attraction between galaxies.

3. **Quantum Effects**:
    ○ Quantum mechanics and the uncertainty principle introduce fundamental uncertainties into the behavior of matter and energy on small scales.
    ○ These quantum effects could potentially prevent a complete collapse of the universe into a singularity, leading to alternative scenarios such as a "Big Rip" or a "Big Freeze."

4. **Unknown Physics**:
    ○ Our understanding of the fundamental forces and properties of the universe is incomplete, particularly at the extreme conditions near a singularity.
    ○ Without a complete theory of quantum gravity or a unified theory of fundamental forces, it is challenging to predict the behavior of the universe under such extreme conditions.

Future Directions and Speculations

While the Big Crunch theory has fallen out of favor among cosmologists, it continues to inspire speculative thinking and theoretical exploration:

1. **Modified Cosmological Models**:

- Some modified cosmological models incorporate elements of the Big Crunch scenario, such as cyclic universe models or brane cosmology in string theory.
    - These models propose alternative mechanisms for cyclic behavior or collapse without violating observational constraints.
2. **Quantum Gravity Effects**:
    - Investigating the behavior of matter and energy near a singularity requires a theory of quantum gravity that unifies quantum mechanics and general relativity.
    - Research in quantum gravity seeks to understand the nature of space-time at the Planck scale and its implications for cosmology.
3. **Multiverse Hypotheses**:
    - Some multiverse hypotheses suggest that our universe is just one of many in a larger ensemble of universes with different properties and fates.
    - In this context, the Big Crunch may be realized in some universes while others continue to expand indefinitely.

Conclusion

The Big Crunch theory, while an intriguing concept, remains speculative and highly uncertain in light of current observational data and theoretical developments in cosmology. While it offers a fascinating glimpse into the potential fate of the universe, the overwhelming evidence supports the idea of a continued expansion driven by dark energy. Nonetheless, the Big Crunch remains an important theoretical construct that has shaped our understanding of cosmology and the dynamic evolution of the universe. As research and technological advancements continue to push the boundaries of our knowledge, further insights into the ultimate fate of the cosmos may emerge, shedding light on the profound questions surrounding the nature of existence and the destiny of the universe. Whether the universe experiences a Big Crunch, a Big Freeze, or another fate altogether, the quest to unravel the mysteries of cosmic evolution remains one of the most profound endeavors in human exploration and discovery.

Chapter 20: Many Worlds Interpretation

The Many Worlds Interpretation (MWI) is a prominent and controversial theory in quantum mechanics that proposes a radical solution to the measurement problem and the nature of reality itself. Developed in the 1950s by physicist Hugh Everett III, MWI posits that every quantum event results in the branching of the universe into multiple, parallel realities, each representing a different outcome of the event. In this detailed exploration, we will delve into the origins, principles, implications, criticisms, and ongoing debates surrounding the Many Worlds Interpretation.

Origins and Development

The roots of the Many Worlds Interpretation can be traced back to the early years of quantum mechanics, when physicists struggled to make sense of the theory's strange and counterintuitive implications. In 1957, Hugh Everett III, a graduate student at Princeton University, proposed a radical new interpretation of quantum mechanics in his doctoral dissertation titled "The Theory of the Universal Wave Function."

Everett's interpretation departed from the prevailing Copenhagen Interpretation, which held that quantum systems exist in a state of superposition until they are observed, at which point they "collapse" into a definite state. Instead, Everett argued that the wave function of the universe never collapses but continues to evolve according to the Schrödinger equation, resulting in a proliferation of parallel realities or "worlds."

Principles of the Many Worlds Interpretation

The Many Worlds Interpretation is grounded in several key principles that distinguish it from other interpretations of quantum mechanics:

1. **Universal Wave Function**:
    - According to MWI, the universe is described by a single, evolving wave function, often referred to as the "universal wave function" or the "wave function of the universe."
    - This wave function contains all possible states of the universe, with each state corresponding to a different configuration of particles and fields.
2. **Branching of Realities**:
    - Whenever a quantum event occurs, such as the measurement of a particle's position or momentum, the universe splits into multiple branches, each corresponding to a different outcome of the event.
    - These branches represent parallel realities that coexist alongside our own, with each reality following its own trajectory in the space of all possible states.
3. **No Collapse of the Wave Function**:
    - Unlike other interpretations of quantum mechanics, MWI rejects the idea of wave function collapse.
    - Instead, the wave function of the universe continues to evolve deterministically according to the Schrödinger equation, with no discontinuous jumps or collapses.
4. **Probabilistic Predictions**:
    - While MWI eliminates the need for wave function collapse, it preserves the probabilistic nature of quantum mechanics.
    - The probabilities of different outcomes are encoded in the relative amplitudes of the branches of the wave function, with more probable outcomes corresponding to branches with higher amplitudes.

Implications of the Many Worlds Interpretation

The Many Worlds Interpretation has far-reaching implications for our understanding of quantum mechanics, the nature of reality, and the concept of observation:

1. **Resolution of the Measurement Problem**:
    -   One of the primary motivations for MWI is its elegant solution to the measurement problem in quantum mechanics.
    -   In MWI, the act of measurement is simply one branch of the universal wave function interacting with another, rather than a collapse of the wave function.

2. **Quantum Superposition and Entanglement**:
    -   MWI provides a natural explanation for phenomena such as quantum superposition and entanglement.
    -   Rather than being mysterious or paradoxical, these phenomena arise naturally from the evolution of the universal wave function and the branching of realities.

3. **Reality of Parallel Worlds**:
    -   According to MWI, parallel worlds are not merely hypothetical constructs but concrete physical realities that exist alongside our own.
    -   Each world is as real and valid as our own, with its own inhabitants, histories, and futures.

4. **Cosmological Implications**:
    -   MWI has implications for cosmology and the structure of the universe on cosmic scales.
    -   It suggests that the multiverse is not a collection of disconnected universes but a single, vast quantum system comprising an infinite number of parallel

realities.

Criticisms and Controversies

Despite its theoretical elegance, the Many Worlds Interpretation has faced criticism and controversy within the physics community:

1. **Ontological Complexity**:
    - Critics argue that MWI introduces unnecessary ontological complexity by postulating the existence of an infinite number of parallel worlds.
    - They question the scientific value of positing the reality of unobservable worlds beyond our own.
2. **Probability Problem**:
    - MWI preserves the probabilistic predictions of quantum mechanics but raises questions about how probabilities are assigned to different branches of the wave function.
    - Critics argue that MWI lacks a satisfactory explanation for the emergence of probabilities and the Born rule.
3. **Observational Consequences**:
    - MWI makes the striking prediction that observers should experience "quantum immortality," where they continually find themselves in branches of the wave function where they survive.
    - However, this prediction is difficult to test empirically and raises philosophical questions about the nature of personal identity.
4. **Interpretational Freedom**:
    - Some critics argue that MWI is not strictly a scientific theory but an interpretational framework that is compatible with a wide range of experimental

outcomes.

- ○ As such, they question whether MWI can be considered a genuine scientific hypothesis.

Ongoing Research and Future Directions

Research on the Many Worlds Interpretation continues to be a lively and active area of inquiry, with ongoing efforts focused on several fronts:

1. **Foundational Questions**:
    - ○ Physicists continue to explore the conceptual foundations of MWI, seeking to address criticisms and clarify its implications for our understanding of quantum mechanics and reality.
2. **Experimental Tests**:
    - ○ Efforts are underway to develop experimental tests that could potentially distinguish between MWI and alternative interpretations of quantum mechanics.
    - ○ These tests may involve exploring the consequences of MWI for phenomena such as quantum superposition, entanglement, and measurement.
3. **Cosmological Signatures**:
    - ○ Cosmologists are investigating whether MWI could leave observable signatures on the cosmic microwave background or other large-scale structures of the universe.
    - ○ Discovering such signatures would provide indirect evidence in support of MWI and its implications for cosmology.
4. **Philosophical Inquiry**:
    - ○ Philosophers continue to explore the philosophical implications of MWI for our understanding of

reality, consciousness, and the nature of existence.

- ○ Debates about the metaphysical status of parallel worlds and the role of observation in quantum mechanics remain topics of active discussion.

Conclusion

The Many Worlds Interpretation represents a bold and provocative attempt to grapple with the profound mysteries of quantum mechanics and the nature of reality. While controversial and speculative, MWI offers a compelling framework for understanding the quantum world and resolving longstanding puzzles in physics.

As research and debate on MWI continue, physicists and philosophers alike are confronted with profound questions about the nature of existence, the structure of the universe, and the limits of human knowledge. Whether MWI ultimately proves to be the correct interpretation of quantum mechanics or not, its exploration has deepened our understanding of the quantum realm and expanded the horizons of scientific inquiry.

Chapter 21: Omega Point Theory

The Omega Point Theory, proposed by physicist and mathematician Frank J. Tipler in the late 20th century, is a speculative hypothesis that combines concepts from physics, cosmology, and theology to envision the ultimate fate and purpose of the universe. Tipler's theory posits that the universe is evolving toward a final state of infinite information processing and intelligence, culminating in a singularity known as the Omega Point. In this comprehensive exploration, we will delve into the origins, principles, implications, criticisms, and ongoing debates surrounding the Omega Point Theory.

Origins and Development

The Omega Point Theory emerged from Tipler's research in the fields of physics, cosmology, and the philosophy of science. In his 1994 book "The Physics of Immortality," Tipler outlined his vision of the universe's ultimate destiny and the role of intelligent life in shaping its future. Drawing on concepts from physics, including general relativity and quantum mechanics, as well as theological ideas, Tipler proposed a grand narrative of cosmic evolution leading to a final singularity of infinite intelligence and power.

Principles of the Omega Point Theory

The Omega Point Theory is founded on several key principles that shape its vision of the universe's ultimate fate and purpose:

1. **Cosmic Evolution:**
    - Tipler envisions the universe as a vast, evolving system that undergoes a series of stages from the Big Bang to the Omega Point.
    - Over billions of years, the universe evolves through

processes of star formation, galaxy formation, and the emergence of life.

2. **Technological Progress:**
    ◦ According to Tipler, intelligent life plays a central role in the universe's evolution, driving technological progress and accelerating the process of complexity and organization.
    ◦ Technological advancements, such as artificial intelligence and space colonization, contribute to the universe's transformation toward the Omega Point.

3. **Ultimate Destiny:**
    ◦ The Omega Point represents the final state of the universe, characterized by infinite intelligence, energy, and information processing capacity.
    ◦ At the Omega Point, the universe achieves a state of maximum organization and complexity, where all information is processed and stored.

4. **Theological Implications:**
    ◦ Tipler's theory incorporates theological concepts, suggesting that the Omega Point corresponds to the concept of God or the divine in various religious traditions.
    ◦ He argues that the Omega Point possesses attributes traditionally ascribed to God, including omnipotence, omniscience, and omnipresence.

The Omega Point and Technological Singularity

Tipler's Omega Point Theory shares similarities with the concept of the technological singularity, popularized by futurists and transhumanists. The technological singularity refers to a hypothetical future event where artificial intelligence surpasses human intelligence, leading to an exponential growth in technological progress and societal change.

Tipler argues that the Omega Point represents the ultimate realization of the technological singularity, where intelligence reaches its maximum potential and transcends the limitations of biology and physics.

Implications of the Omega Point Theory

The Omega Point Theory has profound implications for our understanding of the universe, consciousness, and the nature of reality:

1. **Purpose and Meaning**:
    - Tipler's theory offers a vision of the universe imbued with purpose and meaning, where intelligent life plays a central role in shaping its ultimate destiny.
    - The Omega Point represents the culmination of cosmic evolution and the fulfillment of the universe's inherent potential.
2. **Transcendence and Immortality**:
    - The Omega Point promises the transcendence of physical limitations and the attainment of immortality for intelligent beings.
    - Tipler suggests that individuals may upload their consciousness into a cosmic computer at the Omega Point, preserving their identities for eternity.
3. **Ethical Imperatives**:
    - Tipler argues that the pursuit of technological progress and the advancement of civilization are ethical imperatives that align with the universe's evolutionary trajectory.
    - He advocates for a global effort to accelerate technological development and ensure the survival and expansion of intelligent life.
4. **Interdisciplinary Synthesis**:

- The Omega Point Theory represents an ambitious attempt to integrate ideas from physics, cosmology, theology, and philosophy into a comprehensive worldview.
- It highlights the interconnectedness of scientific and spiritual perspectives on the nature of reality and human existence.

Criticisms and Challenges

Despite its ambitious scope and grand vision, the Omega Point Theory has faced criticism and skepticism from both within and outside the scientific community:

1. **Scientific Plausibility**:
   - Many physicists and cosmologists view Tipler's theory as speculative and lacking empirical support.
   - Critics argue that the Omega Point relies on speculative extrapolations of current scientific knowledge and overlooks fundamental limitations of physics, such as the second law of thermodynamics.
2. **Theological Interpretation**:
   - The incorporation of theological concepts into Tipler's theory has drawn criticism from both religious and secular perspectives.
   - Some theologians reject the idea of equating the Omega Point with God, viewing it as a reductionist and anthropocentric interpretation of religious concepts.
3. **Ethical Concerns**:
   - Critics raise ethical concerns about the implications of the Omega Point Theory, particularly regarding the potential consequences of technological

transcendence and the pursuit of immortality.

- They caution against the hubris of attempting to control or manipulate the universe's ultimate destiny and question the ethical implications of prioritizing technological progress over other values, such as environmental sustainability and social justice.

1. **Alternative Interpretations**:
   - Some physicists and philosophers propose alternative explanations for the universe's ultimate fate, such as the heat death scenario or the idea of a cyclic universe.
   - These alternative theories offer different perspectives on cosmic evolution and challenge the uniqueness of Tipler's Omega Point hypothesis.

Ongoing Research and Future Directions

Research on the Omega Point Theory continues to inspire debate and exploration across multiple disciplines, with several avenues for future investigation:

1. **Cosmological Modeling**:
   - Physicists and cosmologists continue to develop mathematical models and simulations to explore the implications of the Omega Point Theory within the framework of general relativity and quantum mechanics.
   - These models aim to test the feasibility and plausibility of Tipler's hypothesis and identify potential observational signatures that could support or refute it.
2. **Philosophical Inquiry**:

- Philosophers and theologians engage in interdisciplinary dialogue to examine the philosophical and theological implications of the Omega Point Theory.
  - They explore questions about the nature of consciousness, the meaning of existence, and the relationship between science and spirituality raised by Tipler's hypothesis.

3. **Ethical Reflection:**
  - Ethicists and social scientists investigate the ethical and societal implications of the Omega Point Theory, considering questions about the responsible use of technology, the distribution of benefits and risks, and the impact on human values and identities.
  - They seek to foster informed public discourse and decision-making on issues related to the future of humanity and the cosmos.

4. **Interdisciplinary Collaboration:**
  - Collaboration between scientists, philosophers, theologians, and other scholars facilitates a holistic understanding of the Omega Point Theory and its implications.
  - By bridging disciplinary boundaries and integrating diverse perspectives, researchers can deepen their insights into the nature of reality and the human condition.

Conclusion

The Omega Point Theory represents a bold and provocative attempt to address some of the most profound questions about the nature of the universe and humanity's place within it. While speculative and controversial, Tipler's hypothesis challenges us to reconsider our assumptions about cosmic evolution, consciousness, and the ultimate fate of reality.

As research on the Omega Point Theory progresses, scholars from diverse fields continue to explore its implications, test its predictions, and engage in critical reflection on its philosophical and ethical dimensions. Whether the Omega Point represents a genuine cosmic destiny or a speculative hypothesis remains an open question, but its exploration invites us to contemplate the grandeur and mystery of the universe and our role in shaping its future.

Chapter 22: RNA World Hypothesis

The RNA World Hypothesis is a widely accepted scientific theory that proposes RNA (ribonucleic acid) as the precursor to modern life forms on Earth. It suggests that in the early stages of Earth's history, RNA molecules played a central role in both storing genetic information and catalyzing biochemical reactions necessary for life. In this comprehensive exploration, we will delve into the origins, principles, evidence, implications, criticisms, and ongoing research related to the RNA World Hypothesis.

Origins and Development

The concept of the RNA World Hypothesis emerged in the 1960s and 1970s as scientists sought to understand the origins of life on Earth. The idea gained traction following the discovery of ribozymes, RNA molecules capable of catalyzing chemical reactions, and the recognition of RNA's dual role in genetic storage and enzymatic activity. The RNA World Hypothesis was formally proposed in the 1980s by Walter Gilbert, Carl Woese, and others as a plausible explanation for the transition from prebiotic chemistry to the emergence of life.

Principles of the RNA World Hypothesis

The RNA World Hypothesis is grounded in several key principles that underpin its central thesis:

1. **RNA as Genetic Material**:
    - RNA is capable of both storing genetic information, like DNA, and catalyzing biochemical reactions, like proteins.
    - This versatility suggests that RNA may have served as the primary genetic material in early life forms,

predating the emergence of DNA and proteins.

2. **Prebiotic Synthesis**:
    ◦ The RNA World Hypothesis posits that RNA molecules could have arisen through prebiotic synthesis from simpler organic molecules present on early Earth.
    ◦ Experiments have demonstrated that RNA nucleotides, the building blocks of RNA, can form spontaneously under conditions simulating early Earth's environment.

3. **RNA Catalysis**:
    ◦ Ribozymes, RNA molecules with catalytic activity, are capable of facilitating a wide range of biochemical reactions, including self-replication and RNA cleavage.
    ◦ This catalytic activity suggests that RNA could have played a crucial role in driving prebiotic chemistry and the emergence of early life forms.

4. **Transition to DNA and Proteins**:
    ◦ Over time, the RNA World hypothesis proposes that RNA gave rise to more complex molecules, such as DNA and proteins, through processes of molecular evolution and natural selection.
    ◦ DNA eventually replaced RNA as the primary genetic material, while proteins assumed the role of enzymatic catalysts, leading to the emergence of modern cellular life.

Evidence for the RNA World Hypothesis

While direct evidence for the RNA World Hypothesis is challenging to obtain due to the immense timescales involved and the lack of fossil records of ancient RNA molecules, several lines of evidence support its plausibility:

1. **Ribozyme Activity:**
    - Experimental studies have demonstrated that RNA molecules can catalyze a variety of chemical reactions, including RNA cleavage, RNA ligation, and peptide bond formation.
    - These ribozymes provide evidence for the catalytic potential of RNA and its ability to perform essential biochemical functions.

2. **RNA Self-Replication:**
    - RNA molecules are capable of self-replication through processes such as template-directed RNA polymerization.
    - Laboratory experiments have shown that short RNA sequences can catalyze their own replication in a process akin to early forms of RNA-based reproduction.

3. **Prebiotic Synthesis:**
    - Studies of prebiotic chemistry have demonstrated that RNA nucleotides and related molecules can form spontaneously under conditions resembling those of early Earth.
    - Experiments simulating the conditions of primordial Earth have produced RNA precursors, such as ribose and nucleobases, suggesting that the synthesis of RNA molecules was feasible in prebiotic environments.

4. **Genetic Fossils:**
    - Some remnants of the RNA World may still exist in modern organisms in the form of RNA-based

enzymes known as ribonucleoproteins (RNPs) or ribonucleoprotein complexes.

- ○ These molecular fossils provide clues to the early evolution of life and the transition from RNA-based to DNA-based genetic systems.

Implications of the RNA World Hypothesis

The RNA World Hypothesis has profound implications for our understanding of the origin and evolution of life on Earth and beyond:

1. **Origin of Life**:
    - ○ The RNA World Hypothesis provides a plausible explanation for how the complex biochemistry of life could have arisen from simpler molecular precursors on early Earth.
    - ○ By elucidating the role of RNA in both genetic information storage and enzymatic catalysis, the hypothesis offers insights into the transition from non-living to living matter.

2. **Molecular Evolution**:
    - ○ RNA-based life forms likely underwent processes of molecular evolution, leading to the emergence of more complex molecules, such as DNA and proteins.
    - ○ Understanding the mechanisms of RNA replication, mutation, and selection sheds light on the evolutionary processes that gave rise to modern life forms.

3. **Astrobiology**:
    - ○ The RNA World Hypothesis has implications for the search for life beyond Earth, as RNA-based biochemistry may be a common feature of life in the universe.

- Astrobiologists study the conditions on other planets and moons to assess the potential for RNA-based life forms to emerge elsewhere in the cosmos.

4. **Biotechnology and Medicine**:
   - Insights gained from the RNA World Hypothesis have practical applications in biotechnology and medicine, including the design of RNA-based therapeutics and the engineering of ribozymes for biocatalysis.
   - RNA-based technologies hold promise for treating genetic diseases, targeting specific genes, and developing novel drugs and vaccines.

Criticisms and Challenges

While widely accepted within the scientific community, the RNA World Hypothesis is not without its critics and challenges:

1. **Complexity of RNA Synthesis**:
   - Critics point out that the synthesis of RNA from prebiotic molecules under early Earth conditions remains a significant challenge. While laboratory experiments have demonstrated the formation of RNA precursors under simulated prebiotic conditions, replicating the full complexity of RNA synthesis and assembly in a primordial environment is difficult.

1. **Alternative Hypotheses**:
   - Some scientists propose alternative hypotheses for the origin of life, such as the metabolism-first or lipid-world scenarios, which emphasize different aspects of early biochemistry.

- These alternative hypotheses suggest that life may have arisen through the self-organization of simple chemical systems without the need for RNA-based genetic replication.

2. **RNA Stability**:
   - RNA is less stable than DNA and prone to degradation under harsh environmental conditions, such as high temperatures or UV radiation.
   - Critics question whether RNA molecules could have persisted long enough on early Earth to facilitate the emergence of life without the protective mechanisms of cellular organisms.

3. **Role of Minerals and Cofactors**:
   - Recent research suggests that minerals and cofactors may have played a crucial role in facilitating RNA synthesis and catalysis in prebiotic environments.
   - Understanding the interplay between RNA molecules and mineral surfaces could provide new insights into the early stages of life's emergence.

Ongoing Research and Future Directions

Research on the RNA World Hypothesis continues to advance our understanding of the origins of life and the role of RNA in early biochemistry. Several avenues for future investigation include:

1. **Experimental Studies**:
   - Scientists conduct experiments to simulate prebiotic conditions and investigate the synthesis, replication, and catalytic activity of RNA molecules.
   - These experiments aim to elucidate the mechanisms by which RNA could have emerged and functioned as the primary genetic material in early life forms.

2. **Molecular Evolution:**
    ◦ Research on molecular evolution seeks to understand how RNA-based life forms transitioned to more complex genetic systems based on DNA and proteins.
    ◦ By studying the evolutionary trajectories of RNA molecules and their interactions with other biomolecules, scientists can reconstruct the early history of life on Earth.
3. **Astrobiology and Exoplanets:**
    ◦ Astrobiologists explore the potential for RNA-based life to exist on other planets and moons within our solar system and beyond.
    ◦ The discovery of exoplanets with environments conducive to RNA synthesis and stability could inform our understanding of the prevalence and diversity of life in the universe.
4. **Origin of Genetic Code:**
    ◦ Scientists investigate the origins of the genetic code and the translation machinery that enables the synthesis of proteins from RNA templates.
    ◦ Understanding how the genetic code evolved from RNA sequences to amino acid sequences could provide insights into the early evolution of cellular life.

Conclusion

The RNA World Hypothesis offers a compelling explanation for the origins of life on Earth and the role of RNA in the transition from non-living to living matter. While still the subject of ongoing research and debate, the hypothesis has provided a framework for understanding the chemical and evolutionary processes that led to the emergence of modern life forms.

By elucidating the properties and functions of RNA molecules in early biochemistry, scientists are gaining new insights into the fundamental processes that underlie the diversity and complexity of life on Earth and the potential for life to exist elsewhere in the universe. As research on the RNA World Hypothesis continues, it promises to deepen our understanding of life's origins and the interconnectedness of all living organisms on our planet and beyond.

Chapter 23: String Theory

String Theory is a theoretical framework in physics that aims to provide a unified description of the fundamental forces and particles in the universe. It proposes that the fundamental building blocks of reality are not point-like particles but rather one-dimensional "strings" or loops of energy. These strings vibrate at different frequencies, giving rise to the diverse particles and forces observed in nature. In this comprehensive exploration, we will delve into the origins, principles, developments, implications, criticisms, and ongoing research related to String Theory.

Origins and Development

String Theory emerged in the late 20th century as physicists sought to reconcile quantum mechanics, which describes the behavior of particles on microscopic scales, with general relativity, which describes the behavior of gravity on cosmic scales. The theory originated from attempts to describe the strong nuclear force, one of the four fundamental forces of nature, in the context of quantum field theory.

The early formulations of String Theory encountered difficulties due to the presence of unwanted massless particles called tachyons. However, in the late 1960s and early 1970s, breakthroughs by physicists Gabriele Veneziano, Leonard Susskind, Holger Bech Nielsen, and others led to the development of a consistent mathematical framework known as bosonic string theory.

Principles of String Theory

String Theory is founded on several key principles that distinguish it from conventional particle physics theories:

1. **Strings as Fundamental Objects**:
    ◦ In String Theory, the fundamental building blocks

of matter and energy are not point-like particles but rather one-dimensional strings or loops.

- These strings can be open or closed and vibrate at different frequencies, giving rise to the diverse particles observed in nature.

2. **Quantum Mechanics and General Relativity**:

- String Theory aims to unify quantum mechanics, which governs the behavior of particles on small scales, with general relativity, which describes the behavior of gravity on large scales.

- By treating strings as extended objects rather than point particles, String Theory seeks to overcome the mathematical difficulties encountered in attempts to quantize gravity.

3. **Extra Dimensions**:

- String Theory posits the existence of extra spatial dimensions beyond the familiar three dimensions of space and one dimension of time.

- These extra dimensions are compactified or "curled up" at extremely small scales, making them undetectable at macroscopic levels but playing a crucial role in the behavior of strings.

4. **Supersymmetry**:

- Many versions of String Theory incorporate supersymmetry, a theoretical framework that postulates a symmetry between particles with integer spin (bosons) and particles with half-integer spin (fermions).

- Supersymmetry is believed to provide a solution to certain problems in particle physics and may play a role in unifying the fundamental forces of nature.

Developments in String Theory

String Theory has undergone significant developments since its inception, leading to several distinct formulations and branches:

1. **Bosonic String Theory**:
    - The earliest formulation of String Theory, known as bosonic string theory, describes only the vibrational modes of strings without considering fermionic particles.
    - Bosonic string theory encountered difficulties due to the presence of tachyons and was later superseded by more advanced formulations.
2. **Superstring Theory**:
    - Superstring theory incorporates supersymmetry and describes strings that can have both bosonic and fermionic properties.
    - Superstring theory is the most widely studied version of String Theory and forms the basis for many of its developments.
3. **M-Theory**:
    - M-Theory is a hypothetical framework that unifies various versions of superstring theory, including Type IIA, Type IIB, heterotic $SO(32)$, and heterotic $E8 \times E8$ string theories.
    - M-Theory postulates the existence of higher-dimensional objects called membranes or "branes" and provides a more comprehensive description of the universe's fundamental structure.
4. **String Compactification**:
    - String compactification is a process by which the extra dimensions of String Theory are curled up or compactified to produce a four-dimensional

spacetime resembling our observable universe.
- Different compactification schemes lead to distinct physical phenomena and may provide explanations for the observed properties of particle physics and cosmology.

Implications of String Theory

String Theory has far-reaching implications for our understanding of the universe at both the microscopic and cosmic scales:

1. **Unification of Fundamental Forces**:
   - String Theory offers the potential to unify the four fundamental forces of nature—gravity, electromagnetism, the weak nuclear force, and the strong nuclear force—into a single theoretical framework.
   - By describing all interactions in terms of vibrating strings, String Theory aims to provide a unified description of the universe's fundamental constituents and interactions.
2. **Resolution of Quantum Gravity**:
   - One of the primary motivations for String Theory is its potential to resolve the long-standing problem of quantum gravity, which arises when attempting to reconcile quantum mechanics with general relativity.
   - By treating gravity as a fundamental force mediated by the exchange of graviton particles, String Theory offers a consistent framework for understanding the behavior of gravity on both microscopic and cosmic scales.
3. **Cosmological Implications**:
   - String Theory has implications for cosmology, the

study of the origin and evolution of the universe.

- Some versions of String Theory propose the existence of multiple universes or "braneworlds" that coexist within a higher-dimensional space, leading to the possibility of a multiverse.

4. **New Physics Beyond the Standard Model**:
    - String Theory predicts the existence of new particles, interactions, and phenomena beyond those described by the Standard Model of particle physics.
    - Experimental tests of String Theory could potentially reveal signatures of supersymmetry, extra dimensions, and other features predicted by the theory.

Criticisms and Challenges

Despite its theoretical elegance and potential, String Theory faces several criticisms and challenges:

1. **Lack of Experimental Evidence**:
    - String Theory has yet to make definitive predictions that can be tested through experimental observation.
    - Critics argue that the theory's reliance on compactified extra dimensions and supersymmetry may make it difficult to confirm or refute using current experimental techniques.

2. **Complexity and Mathematical Rigor**:
    - String Theory is highly complex mathematically and requires advanced techniques from geometry, topology, and quantum field theory.
    - Some physicists question whether the theory is mathematically rigorous and whether it can be formulated in a way that is internally consistent and

predictive.

3. **String Landscape**:
    - String Theory predicts a vast landscape of possible solutions, each corresponding to a different configuration of extra dimensions and physical constants.
    - Critics argue that this "string landscape" makes the theory highly speculative and raises questions about its explanatory power and predictive capability.
4. **Alternative Approaches**:
    - Some physicists advocate for alternative approaches to quantum gravity and fundamental physics, such as loop quantum gravity, causal dynamical triangulation, and emergent gravity.
    - These approaches offer different perspectives on the nature of spacetime, quantum mechanics, and the fundamental forces of nature.

Ongoing Research and Future Directions

Research on String Theory continues to be a vibrant and active area of investigation, with ongoing efforts focused on several fronts:

1. **Mathematical Formulation**:
    - Physicists work to refine the mathematical formulation of String Theory and develop new techniques for studying its properties and predictions.
    - Advances in algebraic geometry, string field theory, and conformal field theory contribute to our understanding of the mathematical structures underlying String Theory.

1. **Experimental Testing**:
    - Efforts are underway to explore potential experimental signatures of String Theory, such as the detection of supersymmetric particles at high-energy particle colliders like the Large Hadron Collider (LHC).
    - Experimental tests of String Theory may also involve observations of cosmic phenomena, such as gravitational waves, cosmic microwave background radiation, and the distribution of galaxies and dark matter.
2. **String Phenomenology**:
    - String phenomenology focuses on deriving testable predictions from String Theory that are consistent with observations of particle physics and cosmology.
    - Researchers study the implications of String Theory for the properties of known particles, the existence of new particles, the behavior of extra dimensions, and other observable phenomena.
3. **Cosmological Consequences**:
    - Cosmologists explore the cosmological implications of String Theory, including the possibility of inflationary scenarios, cosmic strings, and other features of the early universe.
    - Observational tests of these cosmological predictions could provide insights into the validity of String Theory and its ability to describe the universe on cosmological scales.

Conclusion

String Theory represents a bold and ambitious attempt to formulate a unified theory of fundamental physics that encompasses all known forces and particles in the universe. Despite facing significant challenges and criticisms, String Theory has sparked a revolution in theoretical physics and led to new insights into the nature of spacetime, quantum mechanics, and the fundamental structure of reality.

As research on String Theory continues, physicists explore its mathematical intricacies, experimental predictions, and cosmological implications in search of a deeper understanding of the universe's fundamental laws. Whether String Theory ultimately proves to be the correct description of reality remains an open question, but its pursuit has already reshaped our conception of the cosmos and expanded the boundaries of human knowledge.

Chapter 24: Endosymbiotic Theory

The Endosymbiotic Theory is a scientific hypothesis that proposes the origin of eukaryotic cells—organisms with complex cell structures containing a nucleus and other membrane-bound organelles—through a symbiotic relationship between different prokaryotic organisms. This theory suggests that certain organelles found in eukaryotic cells, such as mitochondria and chloroplasts, were once free-living bacteria that were engulfed by a host cell but, instead of being digested, formed a symbiotic relationship, eventually evolving into the complex cellular structures we observe today. In this detailed exploration, we will delve into the historical background, principles, evidence, implications, criticisms, and ongoing research related to the Endosymbiotic Theory.

Historical Background

The idea of endosymbiosis has its roots in the early 20th century, with various scientists proposing hypotheses about the origin of eukaryotic cells. However, it wasn't until the 1960s and 1970s that the Endosymbiotic Theory gained widespread recognition and support. The theory was popularized and refined by Lynn Margulis, an American biologist, who synthesized existing evidence and proposed a comprehensive framework for understanding the evolutionary origin of eukaryotic cells through symbiosis.

Principles of the Endosymbiotic Theory

The Endosymbiotic Theory is based on several key principles that outline the process by which eukaryotic cells may have evolved from symbiotic relationships between different prokaryotic organisms:

1. **Endosymbiosis**:
    ◦ Endosymbiosis refers to a mutually beneficial

relationship between two different species, where one organism lives within the body or cells of another organism.

- According to the Endosymbiotic Theory, eukaryotic cells originated from endosymbiotic relationships between prokaryotic cells, with one cell engulfing another and forming a symbiotic partnership.

2. **Mitochondria and Chloroplasts**:
   - The Endosymbiotic Theory specifically focuses on the origin of mitochondria and chloroplasts, two essential organelles found in eukaryotic cells.
   - Mitochondria are responsible for cellular respiration and energy production, while chloroplasts are involved in photosynthesis in plant cells.

3. **Prokaryotic Ancestry**:
   - Both mitochondria and chloroplasts share similarities with modern-day bacteria, leading to the hypothesis that they were once free-living prokaryotic organisms.
   - Mitochondria are thought to have evolved from aerobic bacteria capable of cellular respiration, while chloroplasts likely originated from photosynthetic cyanobacteria.

4. **Genetic Evidence**:
   - One of the key pieces of evidence supporting the Endosymbiotic Theory is the presence of circular DNA genomes within mitochondria and chloroplasts, similar to those found in bacteria.
   - These organelles also replicate independently of the host cell's nucleus, further suggesting their bacterial origin.

Evidence for the Endosymbiotic Theory

The Endosymbiotic Theory is supported by a range of evidence from multiple scientific disciplines, including molecular biology, genetics, comparative genomics, and paleontology:

1. **Structural and Functional Similarities**:
    - Mitochondria and chloroplasts share structural and functional similarities with bacteria. For example, both organelles have a double membrane structure reminiscent of bacterial cell walls.
    - Additionally, mitochondria and chloroplasts contain ribosomes and can synthesize proteins, similar to bacteria.
2. **Genetic Evidence**:
    - Comparative genomics has revealed striking similarities between the genomes of mitochondria, chloroplasts, and free-living bacteria.
    - Analysis of mitochondrial and chloroplast DNA has identified genes encoding proteins involved in energy metabolism and photosynthesis, which are homologous to those found in bacteria.
3. **Endosymbiotic Relationships in Nature**:
    - Endosymbiotic relationships between different organisms are common in nature and provide analogies for the origin of mitochondria and chloroplasts.
    - Examples include symbiotic relationships between certain insects and bacteria that provide essential nutrients or between corals and photosynthetic algae living within their tissues.
4. **Fossil Evidence**:
    - While direct fossil evidence of endosymbiotic events

is scarce due to the microscopic size of mitochondria and chloroplasts, there are indirect clues supporting the theory.

- Fossilized remains of cyanobacteria and aerobic bacteria from the Precambrian era provide insights into the types of organisms that existed during the period when eukaryotic cells are believed to have originated.

Implications of the Endosymbiotic Theory

The Endosymbiotic Theory has far-reaching implications for our understanding of the evolution of life on Earth and the diversity of cellular organisms:

1. **Origin of Eukaryotic Cells**:
   - The Endosymbiotic Theory provides a plausible explanation for the origin of eukaryotic cells, which represent a significant evolutionary leap from prokaryotic cells.
   - By proposing that mitochondria and chloroplasts were once independent bacteria that formed symbiotic relationships with host cells, the theory accounts for the complexity and diversity of eukaryotic life.
2. **Evolution of Complexity**:
   - Endosymbiosis represents a key mechanism for the evolution of complexity in living organisms, allowing for the acquisition of new metabolic capabilities and adaptive traits through symbiotic relationships.
   - The incorporation of mitochondria and chloroplasts into eukaryotic cells facilitated the development of multicellular organisms and the colonization of

diverse ecological niches.

3. **Horizontal Gene Transfer**:
   ○ The Endosymbiotic Theory highlights the importance of horizontal gene transfer—the transfer of genetic material between different organisms—in the evolution of cellular life.
   ○ Mitochondria and chloroplasts retain their own genomes, which have undergone gene transfer events with the nuclear genome of the host cell over millions of years of evolution.

4. **Ecological Interactions**:
   ○ Endosymbiotic relationships continue to play crucial roles in ecological interactions and nutrient cycling in modern ecosystems.
   ○ Symbiotic relationships between organisms, such as those between plants and mycorrhizal fungi or between humans and gut bacteria, demonstrate the ongoing significance of symbiosis in shaping biological communities.

Criticisms and Challenges

While widely accepted within the scientific community, the Endosymbiotic Theory is not without its criticisms and challenges:

1. **Origin of Mitochondria and Chloroplasts**:
   ○ The exact mechanisms by which mitochondria and chloroplasts were engulfed by host cells and established stable symbiotic relationships remain speculative.
   ○ Critics question how these organelles avoided digestion by the host cell and acquired mechanisms for replication and division within the eukaryotic

cell.

2. **Alternative Hypotheses**:
   - Some scientists propose alternative hypotheses for the origin of eukaryotic cells, such as the phagotrophic origin hypothesis or the hydrogen hypothesis.
   - These alternative hypotheses emphasize different mechanisms for the origin of cellular complexity and the acquisition of organelles.

3. **Genomic Integration**:
   - The process of genomic integration between organelles and the nuclear genome of the host cell is complex and may involve conflicts between different genetic systems.
   - Understanding the dynamics of gene transfer and genomic integration in endosymbiotic relationships remains an area of active research.

4. **Paleontological Evidence**:
   - Direct fossil evidence of endosymbiotic events is limited, making it challenging to draw definitive conclusions about the timing and mechanisms of organelle acquisition by eukaryotic cells.
   - Fossilized remains of ancient bacteria provide indirect evidence supporting the Endosymbiotic Theory, but the microscopic nature of mitochondria and chloroplasts makes their fossilization improbable.
   - Paleontologists continue to search for additional fossil evidence that could shed light on the evolutionary origins of eukaryotic cells and their organelles.

Ongoing Research and Future Directions

Research on the Endosymbiotic Theory remains a vibrant and active area of investigation, with ongoing efforts focused on several fronts:

1. **Molecular and Genetic Studies**:
    - Scientists conduct molecular and genetic studies to explore the evolutionary relationships between eukaryotic cells, mitochondria, chloroplasts, and their prokaryotic relatives.
    - Comparative genomics and phylogenetic analyses provide insights into the evolutionary histories of different organisms and the genetic mechanisms underlying endosymbiotic relationships.
2. **Experimental Approaches**:
    - Experimental approaches, including laboratory studies and computational simulations, are used to investigate the processes of endosymbiosis and organelle evolution.
    - By manipulating genetic and environmental factors, researchers can simulate different scenarios of organelle acquisition and study the consequences for cellular function and evolution.
3. **Paleontological Discoveries**:
    - Paleontologists continue to search for fossil evidence of ancient microorganisms and early eukaryotic cells that could provide insights into the timing and nature of endosymbiotic events.
    - Advances in imaging techniques and analytical methods may reveal previously undiscovered fossil deposits containing evidence relevant to the Endosymbiotic Theory.
4. **Interdisciplinary Collaboration**:

- ◦ Collaboration between researchers from diverse fields, including microbiology, evolutionary biology, paleontology, and bioinformatics, enhances our understanding of endosymbiosis and organelle evolution.
- ◦ Interdisciplinary approaches facilitate the integration of molecular, genetic, physiological, and ecological data to reconstruct the evolutionary history of eukaryotic cells.

Conclusion

The Endosymbiotic Theory represents a groundbreaking hypothesis that has transformed our understanding of the origin and evolution of eukaryotic cells. By proposing that mitochondria and chloroplasts were once free-living bacteria that formed symbiotic relationships with host cells, the theory provides a compelling explanation for the complexity and diversity of life on Earth.

While the Endosymbiotic Theory is supported by a wealth of evidence from molecular biology, genetics, comparative genomics, and paleontology, it also faces challenges and unanswered questions. Ongoing research continues to refine our understanding of endosymbiosis, organelle evolution, and the mechanisms driving cellular complexity.

As scientists unravel the mysteries of endosymbiosis, they deepen our appreciation for the interconnectedness of all living organisms and the dynamic processes that have shaped life's evolutionary trajectory over billions of years. The Endosymbiotic Theory stands as a testament to the power of symbiosis in driving biological innovation and diversification, illustrating nature's remarkable capacity for adaptation and transformation.

Chapter 25: Electric Universe Theory

The Electric Universe Theory is a cosmological model that challenges conventional views of the universe, proposing that electromagnetic forces play a fundamental role in shaping cosmic structures and phenomena. This theory suggests that electric currents, plasma, and electromagnetic fields are key drivers of celestial processes, including star formation, galaxy dynamics, and the behavior of cosmic bodies. In this comprehensive exploration, we will delve into the origins, principles, evidence, implications, criticisms, and ongoing research related to the Electric Universe Theory.

Origins and Development

The roots of the Electric Universe Theory can be traced back to the early 20th century, with the work of pioneers such as Kristian Birkeland, Hannes Alfvén, and Immanuel Velikovsky. Birkeland's research on the aurora borealis and Alfvén's work on plasma physics laid the groundwork for understanding the role of electricity in space. Velikovsky, though controversial, proposed radical ideas about the role of electromagnetic forces in planetary dynamics in his book "Worlds in Collision."

In the latter half of the 20th century, proponents of the Electric Universe Theory, including plasma physicist Anthony Peratt and engineer Wal Thornhill, expanded upon these ideas, drawing on advances in plasma physics, astronomy, and computer modeling. Their work challenged the dominant paradigm of gravity-centric cosmology, advocating for a more electrically driven view of the universe.

Principles of the Electric Universe Theory

The Electric Universe Theory is based on several key principles that distinguish it from conventional cosmological models:

1. **Electric Plasma**:
   - The Electric Universe Theory posits that plasma, a state of matter consisting of ionized gas, is the dominant form of matter in the universe.
   - Plasma is electrically conductive and responds to electromagnetic forces, giving rise to complex interactions and phenomena on cosmic scales.
2. **Electric Currents**:
   - Electric currents flow through plasma, generating magnetic fields and producing a wide range of electromagnetic effects.
   - These electric currents play a central role in shaping celestial objects and structures, including galaxies, stars, and planetary bodies.
3. **Filamentary Structures**:
   - The Electric Universe Theory suggests that the universe is composed of filamentary structures, with electric currents threading through vast cosmic webs of plasma.
   - These filaments form interconnected networks that channel energy and matter across different scales, influencing the dynamics of galaxies and galaxy clusters.
4. **Electric Discharges**:
   - Electric discharges, such as cosmic lightning bolts or electrical arcing between celestial bodies, are proposed as mechanisms for powering and shaping various cosmic phenomena.
   - These discharges can produce phenomena observed in space, such as jets emanating from active galactic nuclei, planetary auroras, and transient luminous events.

Evidence for the Electric Universe Theory

Proponents of the Electric Universe Theory cite several lines of evidence to support their model:

1. **Plasma Cosmology**:
    - Observations of cosmic plasma phenomena, such as galactic filaments, cosmic jets, and plasma sheaths around celestial bodies, align with predictions of plasma cosmology.
    - The behavior of plasma in laboratory experiments and computer simulations further supports the role of electric currents and magnetic fields in shaping cosmic structures.
2. **Plasma Redshift**:
    - Plasma redshift, a phenomenon observed in laboratory experiments and predicted by plasma cosmology, may offer an alternative explanation for the redshift of light from distant galaxies.
    - According to the Electric Universe Theory, plasma redshift results from the interaction of light with electric fields in space, rather than the expansion of the universe.
3. **Planetary Features**:
    - The Electric Universe Theory provides explanations for geological and atmospheric features observed on planets and moons, including cratering patterns, magnetic fields, and surface electrical phenomena.
    - Electric discharge machining, rather than gradual erosion by wind and water, is proposed as a mechanism for shaping planetary surfaces.
4. **Solar Activity**:
    - Observations of the Sun's behavior, such as sunspots,

solar flares, and coronal mass ejections, support the idea of electric currents and magnetic fields driving solar activity.

- Electric Universe proponents argue that solar phenomena are better understood as manifestations of electric plasma dynamics rather than solely governed by gravitational processes.

Implications of the Electric Universe Theory

The Electric Universe Theory has significant implications for our understanding of cosmology, astrophysics, and planetary science:

1. **Alternative Cosmology**:
   - The Electric Universe Theory challenges the conventional Big Bang model of cosmology, proposing an alternative view of the universe as an electrically interconnected system.
   - Rather than expanding from a singular event, the universe is seen as a dynamic, self-organizing system driven by electromagnetic forces.

2. **Planetary Dynamics**:
   - Electric Universe principles offer new insights into planetary dynamics, including the formation of craters, valleys, and other surface features through electric discharge machining.
   - Atmospheric phenomena, such as lightning and auroras, are interpreted as manifestations of electrical interactions between planets and their environments.

3. **Galactic Evolution**:
   - Electric Universe concepts provide a framework for understanding the evolution of galaxies, including the formation of spiral arms, galactic jets, and

magnetic fields through plasma dynamics.

- Rather than being shaped solely by gravitational interactions, galaxies are influenced by electric currents and magnetic fields threading through cosmic plasma.

4. **Predictive Power:**
- The Electric Universe Theory offers testable predictions that differ from those of conventional cosmological models, providing opportunities for observational and experimental validation.
- Predictions related to plasma phenomena, electromagnetic effects, and the behavior of celestial bodies can be compared with observational data to evaluate the theory's validity.

Criticisms and Challenges

Despite its proponents' enthusiasm, the Electric Universe Theory faces criticism and skepticism from mainstream scientists and cosmologists:

1. **Lack of Consensus:**
- The Electric Universe Theory remains a fringe idea within the scientific community, with many researchers viewing it as speculative and lacking empirical support.
- Critics argue that the theory's emphasis on electric currents and plasma phenomena overlooks the well-established role of gravity in shaping the cosmos, as supported by extensive observational evidence.

1. **Misinterpretation of Observational Data:**
- Skeptics of the Electric Universe Theory argue that its proponents often misinterpret or cherry-pick

observational data to fit their preconceived notions, rather than following the scientific method of forming hypotheses based on empirical evidence.

- There is concern that confirmation bias may lead Electric Universe proponents to see patterns and correlations where none exist, undermining the credibility of their claims.

2. **Lack of Quantitative Modeling**:
    - Critics contend that the Electric Universe Theory lacks detailed quantitative models and predictive power compared to established cosmological frameworks like general relativity and the Big Bang theory.
    - Without rigorous mathematical formalism and computational simulations, the theory struggles to provide detailed explanations and make accurate predictions about observed phenomena.

3. **Incompatibility with Established Physics**:
    - Some aspects of the Electric Universe Theory appear to contradict well-established principles of physics, such as Maxwell's equations of electromagnetism and Einstein's theory of general relativity.
    - Proponents of the theory argue that it represents a paradigm shift in our understanding of the universe, challenging the orthodox views of gravity-dominated cosmology.

Ongoing Research and Future Directions

While the Electric Universe Theory remains controversial, ongoing research and exploration continue to explore its validity and implications:

1. **Laboratory Experiments**:
    - Scientists conduct laboratory experiments to study plasma physics, electric discharge phenomena, and the behavior of electromagnetic fields in simulated cosmic environments.
    - These experiments aim to validate or refute specific predictions of the Electric Universe Theory and provide empirical evidence for its underlying principles.
2. **Observational Studies**:
    - Astronomers and astrophysicists continue to observe celestial phenomena using ground-based telescopes, space-based observatories, and other instruments.
    - By studying plasma dynamics, electromagnetic radiation, and cosmic structures in detail, researchers seek to discern the role of electricity and magnetism in shaping the universe.
3. **Computer Simulations**:
    - Computational modeling and simulations play a crucial role in testing the predictions of the Electric Universe Theory and exploring its implications for cosmic phenomena.
    - Advanced numerical techniques allow scientists to simulate the behavior of plasma, electric currents, and magnetic fields on cosmic scales, providing insights into their effects on galaxy formation, stellar evolution, and other processes.
4. **Interdisciplinary Collaboration**:
    - Collaboration between researchers from different fields, including physics, astronomy, geology, and engineering, facilitates interdisciplinary investigations into the Electric Universe Theory.

- By combining expertise from diverse disciplines, scientists can address complex questions about the nature of cosmic phenomena and the role of electricity in the universe.

Conclusion

The Electric Universe Theory represents a provocative and unconventional approach to understanding the cosmos, proposing that electromagnetic forces and plasma dynamics play a central role in shaping celestial phenomena. While the theory challenges established paradigms of gravity-dominated cosmology, it remains a subject of debate and skepticism within the scientific community.

As researchers continue to explore the implications of the Electric Universe Theory through laboratory experiments, observational studies, and computational modeling, they seek to unravel the mysteries of cosmic electricity and its influence on the evolution of the universe. Whether the Electric Universe Theory will ultimately revolutionize our understanding of the cosmos or remain a fringe idea on the outskirts of mainstream science remains an open question, but its proponents continue to advocate for its validity and significance in shaping our worldview of the universe.

Chapter 26: Drake Equation

The Drake Equation is a probabilistic formula proposed by astrophysicist Dr. Frank Drake in 1961 to estimate the number of extraterrestrial civilizations in our galaxy with which we might be able to communicate. This equation represents an attempt to address the question of whether intelligent life exists elsewhere in the universe and, if so, how common it might be. In this comprehensive exploration, we'll delve into the historical context, components, implications, criticisms, and current status of the Drake Equation, unraveling its profound implications for our understanding of the cosmos and the search for extraterrestrial intelligence (SETI).

Historical Context:

The Drake Equation emerged during a time of growing interest in the possibility of extraterrestrial life and the exploration of the cosmos. In the early 1960s, as humanity entered the space age and began sending spacecraft beyond Earth's atmosphere, scientists and the public alike pondered the profound question of whether we are alone in the universe.

Dr. Frank Drake, an American astrophysicist and pioneer in the search for extraterrestrial intelligence, formulated the Drake Equation during the first scientific meeting on the search for extraterrestrial intelligence (SETI) at the Green Bank Observatory in West Virginia in 1961. This equation was intended as a framework for discussing the factors that might influence the likelihood of detecting signals from extraterrestrial civilizations.

Components of the Drake Equation:

The Drake Equation is expressed as follows:

$$N = R^* \times f_p \times n_e \times f_i \times f_i \times f_c \times L$$

- N represents the number of civilizations we could communicate with.
- R^* is the average rate of star formation.
- f_p is the fraction of those stars with planets.
- n_e is the average number of planets that could potentially support life.
- f_i is the fraction of planets where life develops.
- f_i is the fraction of planets with intelligent life.
- f_c is the fraction of planets with technology capable of communication.
- L is the lifespan of communicating civilizations.

Implications and Interpretation:

The Drake Equation has profound implications for our understanding of the universe and our place within it:

1. **Cosmic Perspective:** The equation highlights the vast scale and complexity of the cosmos, suggesting that our galaxy may be teeming with potentially habitable planets and intelligent life forms. It invites us to consider our place in the universe and the possibility of sharing it with other civilizations.

2. **Search for Extraterrestrial Intelligence:** The Drake Equation serves as a guiding framework for the search for extraterrestrial intelligence (SETI), informing the design of observational strategies and the interpretation of data from radio telescopes and other instruments. By quantifying the factors that influence the likelihood of detecting signals from extraterrestrial civilizations, the equation helps prioritize target stars and regions for observation.

3. **Astrobiological Exploration:** The equation underscores the interdisciplinary nature of astrobiology, the study of the origin, evolution, and distribution of life in the universe. By considering factors such as planetary habitability, the emergence of life, and the development of intelligence, the Drake Equation highlights the interconnectedness of astronomy, biology, and other scientific disciplines.

4. **Technological Evolution:** The equation prompts us to consider the trajectory of technological evolution and its implications for the development of interstellar communication. Factors such as the duration of detectable signals (L) and the fraction of civilizations capable of interstellar communication (f_c) reflect assumptions about the evolution of technology and the sustainability of advanced civilizations.

Criticisms and Challenges:

Despite its significance, the Drake Equation also faces criticisms and challenges:

1. **Uncertainties and Assumptions:** The equation relies on numerous uncertain parameters and assumptions, including the rate of star formation, the prevalence of planetary systems, and the likelihood of life and intelligence emerging on other worlds. The uncertainties inherent in these factors make it difficult to arrive at a precise estimate of the number of detectable civilizations.

2. **Sampling Bias:** The Drake Equation may be affected by sampling bias, as our observations are limited to a small fraction of the Milky Way galaxy. The equation assumes that our solar system is representative of the broader population of planetary systems, which may not be the case.

3. **Temporal and Spatial Variability:** The factors in the Drake Equation may vary over time and space, leading to fluctuations in the number of detectable civilizations at any given moment. Factors such as the duration of detectable signals (L) and the fraction of civilizations capable of interstellar communication (f_c) may change over cosmic timescales.

4. **Technological Limitations:** The Drake Equation assumes that advanced civilizations use radio or other electromagnetic signals for interstellar communication, which may not be a universal characteristic. Advanced civilizations may employ alternative communication methods or remain undetectable to our current instruments.

Current Status and Future Directions:

The Drake Equation continues to stimulate scientific inquiry and debate, serving as a framework for ongoing research in astrobiology, SETI, and the study of planetary systems:

1. **Astrobiological Exploration:** Advances in exoplanet discovery and characterization have expanded our understanding of planetary systems and their potential for habitability. Future missions, such as the James Webb Space Telescope and the upcoming LUVOIR and HabEx missions, will further our search for potentially habitable exoplanets and signs of life.

2. **Technological Innovation:** The development of new observational techniques and instrumentation, including next-generation radio telescopes, optical telescopes, and space-based observatories, promises to revolutionize our ability to detect and study extraterrestrial civilizations. Breakthrough initiatives such as the Breakthrough Listen project and the Search for Extraterrestrial Intelligence (SETI)

Institute continue to push the boundaries of SETI research.

3. **Interdisciplinary Collaboration:** The Drake Equation encourages interdisciplinary collaboration among astronomers, biologists, chemists, physicists, and other scientists to address fundamental questions about the origin, evolution, and distribution of life in the universe. By integrating insights from diverse fields, researchers can develop more comprehensive models of planetary habitability and the potential for extraterrestrial intelligence.

4. **Philosophical Reflection:** The Drake Equation prompts us to consider broader questions about the nature of life, intelligence, and civilization in the cosmos. Discussions about the implications of contact with extraterrestrial civilizations, the ethical considerations of messaging to other worlds, and the cultural impact of discovering alien life enrich our philosophical reflection on humanity's place in the universe.

In conclusion, the Drake Equation represents a landmark contribution to the field of astrobiology and the search for extraterrestrial intelligence. While it remains a theoretical framework with inherent uncertainties and assumptions, the equation serves as a catalyst for scientific inquiry and exploration, inspiring researchers to investigate the fundamental questions about life's origins and the prevalence of intelligent civilizations in the cosmos. As our understanding of planetary systems, astrobiology, and technology continues to advance, the Drake Equation will remain a touchstone for our quest to unravel the mysteries of the universe and our place within it.

Chapter 27: Bekenstein Bound

The Bekenstein Bound, proposed by physicist Jacob Bekenstein in the early 1970s, is a fundamental principle in theoretical physics that sets an upper limit on the amount of information that can be contained within a finite region of space with a given amount of energy. This concept arises from the intersection of quantum mechanics, general relativity, and thermodynamics, and it has profound implications for our understanding of black holes, entropy, and the nature of information in the universe. In this detailed exploration, we will delve into the historical background, underlying principles, mathematical formulation, implications, and current status of the Bekenstein Bound, unraveling its profound implications for our understanding of the fundamental laws of physics.

Historical Background:

The Bekenstein Bound is named after Jacob Bekenstein, an Israeli-American physicist who developed the concept in the early 1970s while studying the thermodynamics of black holes. Bekenstein's work was motivated by efforts to reconcile the laws of thermodynamics with the properties of black holes, which are objects predicted by general relativity to possess an event horizon from which nothing, not even light, can escape.

Underlying Principles:

The Bekenstein Bound is based on several key principles from quantum mechanics, general relativity, and thermodynamics:

1. **Entropy:** In thermodynamics, entropy is a measure of the disorder or randomness of a system. It is related to the number of microscopic configurations or arrangements of particles

that correspond to a given macroscopic state. Entropy is a fundamental concept in statistical mechanics and plays a central role in understanding the behavior of physical systems.

2. **Black Hole Thermodynamics:** In the 1970s, physicists Stephen Hawking and Roger Penrose independently demonstrated that black holes possess thermodynamic properties, including entropy, temperature, and energy. Hawking's discovery of black hole radiation, now known as Hawking radiation, provided a theoretical basis for understanding the thermodynamic properties of black holes.

3. **Information Theory:** In information theory, information is a measure of the amount of uncertainty or surprise associated with the outcome of an event. It is related to the number of bits or binary digits required to specify the state of a system. Information theory provides a formal framework for understanding the transmission, storage, and processing of information.

Implications:

The Bekenstein Bound has profound implications for our understanding of black holes, entropy, and the nature of information in the universe:

1. **Black Hole Information Paradox:** One of the most significant implications of the Bekenstein Bound is its resolution of the so-called black hole information paradox. According to classical general relativity, information that falls into a black hole is seemingly lost forever, as nothing can escape from within the event horizon. However, the Bekenstein Bound implies that the entropy of a black hole is proportional to its surface area rather than its volume, suggesting that information may be encoded in the surface of

the black hole.

2. **Holographic Principle:** The Bekenstein Bound is closely related to the holographic principle, a conjecture in theoretical physics that suggests that the information content of a three-dimensional region of space can be encoded on a two-dimensional surface surrounding the region. This principle is supported by insights from string theory and black hole physics and provides a unified framework for understanding the quantum nature of gravity.

3. **Quantum Gravity:** The Bekenstein Bound provides important constraints on theories of quantum gravity, which seek to reconcile quantum mechanics with general relativity at the smallest scales. The bound implies a fundamental limit on the amount of information that can be stored or processed within a finite region of space, offering insights into the microscopic structure of spacetime.

4. **Cosmology:** The Bekenstein Bound has implications for cosmology and the study of the early universe. It suggests that the total amount of entropy contained within the observable universe is finite and bounded by its size and energy content. Understanding the implications of the Bekenstein Bound for the cosmic entropy and information content of the universe is an active area of research in cosmology.

Criticisms and Challenges:

While the Bekenstein Bound is widely accepted within the physics community, it is not without its criticisms and challenges:

1. **Quantum Corrections:** The original formulation of the Bekenstein Bound is based on semiclassical arguments and does not include quantum corrections or gravitational effects. Incorporating quantum and gravitational corrections into the

bound is a subject of ongoing research and may lead to refinements of the original inequality.

2. **Violation of Unitarity:** The implications of the Bekenstein Bound for the conservation of quantum information and the unitarity of quantum mechanics remain subjects of debate. Resolving apparent conflicts between the Bekenstein Bound and the principles of quantum mechanics is an active area of research in theoretical physics.

3. **Applicability to Cosmology:** Extending the Bekenstein Bound to cosmological scales and understanding its implications for the entropy and information content of the universe present challenges. Cosmological models and observations may provide insights into the validity and applicability of the bound on the largest scales.

4. **Experimental Tests:** Testing the Bekenstein Bound experimentally poses significant challenges due to the extreme conditions required to probe the quantum and gravitational regimes. Developing experimental techniques and technologies to test the predictions of the bound is an ongoing area of research in experimental physics.

Current Status and Future Directions:

The Bekenstein Bound remains a fundamental principle in theoretical physics, providing important insights into the relationship between entropy, energy, and information in the universe. Future directions for research on the Bekenstein Bound include:

1. **Quantum Gravity:** Investigating the implications of the Bekenstein Bound for theories of quantum gravity, including string theory, loop quantum gravity, and holographic models of spacetime. Understanding the microscopic structure of black holes and the nature of spacetime at the Planck scale is a

central goal of theoretical physics.

2. **Black Hole Physics:** Exploring the connections between the Bekenstein Bound, black hole thermodynamics, and the holographic principle. Investigating the information content of black holes and the resolution of the black hole information paradox remains a topic of active research in theoretical physics.

3. **Cosmology:** Studying the cosmological implications of the Bekenstein Bound for our understanding of the entropy and information content of the universe. Investigating the role of the bound in cosmological models, observational cosmology, and the study of the early universe is an important area of research in cosmology.

4. **Experimental Tests:** Developing experimental techniques and technologies to test the predictions of the Bekenstein Bound in laboratory settings and astrophysical environments. Investigating the validity and applicability of the bound under extreme conditions and exploring its implications for the fundamental laws of physics is a challenging but rewarding endeavor.

In conclusion, the Bekenstein Bound represents a profound and far-reaching principle in theoretical physics, providing important insights into the fundamental nature of entropy, energy, and information in the universe. While the bound poses challenges and raises questions about the nature of spacetime and the laws of quantum mechanics, it also offers a unifying framework for understanding the quantum nature of gravity and the microscopic structure of black holes. As research on the Bekenstein Bound continues to progress, it promises to deepen our understanding of the fundamental laws that govern the cosmos and illuminate the nature of reality at the smallest and largest scales.

Epilogue

As we come to the end of our journey through the bold theories that have shaped our understanding of the universe, we are left with a profound sense of wonder and awe at the boundless mysteries that surround us. From the smallest particles to the vast reaches of space, humanity's quest for knowledge has led us to confront the most fundamental questions of existence, and the theories explored in this book represent our collective effort to unlock the secrets of the cosmos.

Through the visionary insights of groundbreaking thinkers like Albert Einstein, Max Planck, and Stephen Hawking, we have gained new perspectives on the nature of space, time, and reality itself. From the revolutionary principles of relativity and quantum mechanics to the tantalizing possibilities of string theory and the multiverse, the bold theories presented in these pages have expanded the horizons of human knowledge and imagination.

But our journey does not end here. As we reflect on the remarkable achievements of the past, we also look to the future with anticipation and excitement. The quest to understand the universe is an ongoing endeavor, and the challenges that lie ahead are as daunting as they are exhilarating.

In the coming years and decades, new discoveries will continue to reshape our understanding of the cosmos, and bold theories yet to be conceived will inspire future generations to push the boundaries of human knowledge even further. As we stand on the threshold of a new era of scientific exploration, we are reminded of the words of Carl Sagan, who famously said, "Somewhere, something incredible is waiting to be known."

"The Bold Theories" is more than just a collection of ideas—it is a testament to the indomitable spirit of human curiosity and the enduring quest for truth that drives us ever onward. As we close this chapter and look to the future, may we carry with us the spirit of exploration, the courage to question, and the determination to seek answers to the greatest mysteries of the universe.

For in the pursuit of knowledge, we discover not only the secrets of the cosmos but also the boundless potential of the human spirit to illuminate the darkness and unlock the wonders of the unknown.

Thank you for joining us on this extraordinary journey through "The Bold Theories." May it inspire you to continue exploring, questioning, and dreaming of the limitless possibilities that lie ahead.

The End.

www.ingramcontent.com/pod-product-compliance
Lightning Source LLC
Chambersburg PA
CBHW030315160726
47992CB00005B/2011